RADIO BROADCASTING

RADIO BROADCASTING

A History of the Airwaves

GORDON BATHGATE

To Dan,

We all very sincerely wish you well in this new chapter. We will of course miss you greatly and know that you will be a huge success in all that you do in the future.

With our very best Wishes from the LSA team at Callywith

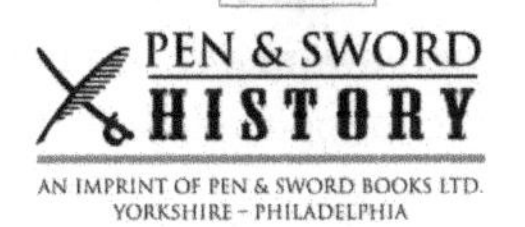

AN IMPRINT OF PEN & SWORD BOOKS LTD.
YORKSHIRE – PHILADELPHIA

First published in Great Britain in 2020 and reprinted in 2021 by
PEN AND SWORD HISTORY
An imprint of
Pen & Sword Books Ltd
Yorkshire – Philadelphia

ISBN 978 1 52676 940 4

A CIP catalogue record for this book is available from the British Library.

Typeset in Times New Roman 11.5/14 by
SJmagic DESIGN SERVICES, India.
Printed and bound in the UK by CPI Group (UK) Ltd, Croydon, CR0 4YY.

Pen & Sword Books Limited incorporates the imprints of Atlas, Archaeology, Aviation, Discovery, Family History, Fiction, History, Maritime, Military, Military Classics, Politics, Select, Transport, True Crime, Air World, Frontline Publishing, Leo Cooper, Remember When, Seaforth Publishing, The Praetorian Press, Wharncliffe Local History, Wharncliffe Transport, Wharncliffe True Crime and White Owl.

For a complete list of Pen & Sword titles please contact
PEN & SWORD BOOKS LIMITED
47 Church Street, Barnsley, South Yorkshire, S70 2AS, England
E-mail: enquiries@pen-and-sword.co.uk
Website: www.pen-and-sword.co.uk

Or
PEN AND SWORD BOOKS
1950 Lawrence Rd, Havertown, PA 19083, USA
E-mail: Uspen-and-sword@casematepublishers.com
Website: www.penandswordbooks.com

Contents

Acknowledgements

This book is dedicated to James Clerk Maxwell, Heinrich Hertz and Guglielmo Marconi. Thanks for inventing this marvellous thing called radio, which has given me a lifetime of pleasure and happiness. Thank you to the team at Pen & Sword Books for their help in the preparation of this book. Finally, I am particularly grateful to my long-suffering wife, Karen 'Oh No! Not another radio' Bathgate. Your patience and understanding regarding my radio obsession is greatly appreciated.

Chapter 1

The Fathers of Radio

On a chilly evening early in 1922, in an old army hut located in a small Essex village, a small, ragtag band of men had gathered. It was St Valentine's Day but romantic thoughts had been pushed to the back of their minds. They were preoccupied with something else; something far more important. As the hands of the timepiece on the wall inched slowly forward to 8pm, they scurried around tweaking and adjusting their equipment. Last-minute checks completed, the men waited eagerly to proceed.

This wasn't a crack team of military personnel about to embark on a secret mission. They were a disparate bunch of engineers brought together by an enterprising Italian called Guglielmo Marconi. Under the auspices of his company, they were about to launch Britain's first regular radio broadcasting service. At the appointed hour the transmitter and aerial crackled into life.

This experimental station, named 2MT and based at Writtle, a small village near Chelmsford, was the first British radio station to make regular entertainment broadcasts. Transmissions emitted from a hut near to the Marconi laboratories. Initially the station only had 200 watts and transmitted on 428 kilohertz on Tuesdays from 8pm to 8.30pm.

The first transmission was far from a triumph. The signal was weak and the sound was muffled. However, things did gradually improve. These early tests consisted mainly of gramophone records but a live concert was broadcast later.

The station, fronted by the eccentric Captain Peter Pendleton Eckersley, a Marconi engineer, was a surprising success. Eckersley, known as 'Captain' or 'PPE' by his friends, was no shrinking violet and never suffered from the dreaded disease of microphone shyness. His light-hearted enthusiasm effervesced across the ether and pervaded each broadcast.

The station was required to read out its allocated call sign '2MT' at regular intervals. The British Army phonetic alphabet was widely used by radio amateurs in 1922; consequently 'M' was 'Emma' and 'T' was 'Toc', so '2MT' became 'Two Emma Toc'. The humorous and glib manner in which Eckersley read out the phonetic version of '2MT' resulted in the station being affectionately known as 'Two Emma Toc'.

PPE liked to experiment with sound and would use whatever was lying around to make unusual noises. He would perform spontaneous comedy sketches and improvise operatic parodies. His regular announcement, 'This is Two Emma Toc, Writtle testing, Writtle testing,' delivered in his jocular style, became extremely well known in a very short space of time.

The medium of radio had been established but its genesis had been a prolonged affair. In the late nineteenth century it was clear to numerous scientists that wireless communication was possible. Various theoretical and experimental advancements led to the development of radio and the communication system we know today. The key invention for the beginning of 'wireless transmission of data using the entire frequency spectrum' was the spark-gap transmitter. These devices served as the transmitters for most wireless telegraphy systems for the first three decades of radio.

During its early development, and long after widespread use of the technology, disputes persisted as to the person who could claim sole credit for the invention of radio. Many experiments were running concurrently and across continents. Some scientific theories were merely notional and later verified as unworkable, but they also helped fuel other ideas that did advance technology. There are several men who have been proclaimed the 'father of radio', but perhaps one simple way to sort out the parentage is to place events in a rough chronological order.

Numerous scientists had posited that electricity and magnetism were linked in some way, but while both are capable of causing attraction and repulsion of objects, they remain distinct effects. In 1802 Gian Domenico Romagnosi proposed the relationship between electric current and magnetism, but his reports were largely ignored.

In 1820 Hans Christian Ørsted publicly conducted an experiment that demonstrated the relationship between electricity and magnetism in a very simple way. He established that a wire carrying a current could deflect a magnetised compass needle. His initial interpretation was that magnetic effects radiate from all sides of a wire carrying an electric current, as do light and heat. Three months later he began more thorough investigations and subsequently published his findings. Øersted's work influenced André-Marie Ampère's theory of electromagnetism.

The British physicist Michael Faraday had discovered the existence of electromagnetic fields in 1845. After becoming interested in science, Faraday began working with Humphrey Davy, the renowned chemist and inventor. Davy gave Faraday a valuable scientific education and also introduced him to important European scientists.

Faraday's greatest contribution to science was in the field of electricity. In 1831 he began a series of experiments in which he discovered electromagnetic induction. Faraday developed the theory that a current flowing in one wire could induce a current in another wire that was not physically connected to the first.

Although Faraday was the first to publish his results, the American scientist Joseph Henry had been working with electromagnetism. Henry had invented a forerunner to the electric doorbell, in particular a bell that could be rung at a distance via an electric wire.

James Clerk Maxwell, the Scottish physicist, was fascinated by Faraday and Henry's work on electromagnetism. He noticed that electrical and magnetic fields could couple together to form electromagnetic waves. Neither an electrical field such as the static which forms when you rub your feet on a carpet, nor a magnetic field like the one that holds a magnet onto a refrigerator will go anywhere on their own account. Nevertheless, Maxwell discovered that a varying magnetic field would induce a varying electric field and vice-versa.

An electromagnetic wave subsists when the changing magnetic field causes another changing electric field, which then causes yet another changing magnetic field, and so on in perpetuity. Unlike a static field, a wave can't exist unless it is moving. Once produced, an electromagnetic wave will carry on forever unless absorbed by matter.

In 1864 Maxwell published his first paper that showed by theoretical reasoning that an electrical disturbance resulting from a change in an electrical quantity, such as voltage or current, should propagate through space at the speed of light. Maxwell finally published this work in his *Treatise on Electricity and Magnetism* in 1873.

In 1866 Dr Mahlon Loomis described a system of signalling by radio. He proposed the theory that the Earth's upper atmosphere was divided into separate concentric layers, and these layers could be tapped by metallic conductors on hills and mountain tops. This was to provide long-distance wireless telegraph and telephone communication, as well as draw electricity down to the Earth's surface.

Dr Loomis claimed to have transmitted signals between two Blue Ridge Mountain tops 22km apart in Virginia, using two kites as antennas. The kites had 180-m-long wires attached to them. Both ends were grounded; one through a galvanometer. When he disconnected and reconnected one end, the amount of current flowing through the other end changed. He therefore claimed to be the first person to achieve wireless, electronic communication. His idea of conductive atmospheric layers has since been discredited.

Mahlon Loomis received a patent for a 'wireless telegraph' in July 1872. This patent claimed to eliminate the overhead wire used by the existing telegraph systems by utilising atmospheric electricity. It didn't contain diagrams or specific methods and didn't refer to, or incorporate, any known scientific theory. It is markedly similar to William Henry Ward's patent that was issued a few months earlier. Neither patent referred to any known scientific theory of electromagnetism and could never have received and transmitted radio waves. It's widely assumed that Loomis exaggerated his achievements to sustain interest in a system that he undoubtedly believed would work.

Towards the end of 1875, while experimenting with the telegraph, Thomas Edison described a phenomenon that he termed 'etheric force', which would later be known as high frequency electromagnetic waves. He announced it to the press on 28 November but cancelled this avenue of research when Elihu Thomson, the engineer and inventor, ridiculed the idea. It was not based on the electromagnetic waves described by Maxwell.

In 1879 David Edward Hughes was the first to claim to have transmitted and received radio waves. He spotted what seemed to be a new phenomenon during his experiments. He realised that sparking in one device could be heard in a separate portable microphone apparatus he had installed nearby. He demonstrated his discovery to the Royal Society in 1880, nine years before electromagnetic radiation was a proven concept. It was most probably radio transmissions, but others convinced Hughes that his discovery was simply induction. Hughes was so demoralised he didn't publish the results of his work and though he continued experimenting with radio, he became diffident and left it to others to document his findings.

In 1884 the Italian Temistocle Calzecchi-Onesti demonstrated a primitive device that would later be developed to become the first practical radio detector. He placed metal filings in a glass box or tube, and made them part of an ordinary electric circuit.

In 1890 Frenchman Edouard Branly demonstrated a much-improved version of Calzecchi-Onesti's device. He called his version a 'radio-conductor' (based on the verb 'to radiate': in Latin 'radius' means 'beam of light'). His device would later be known as a 'coherer'. Branly demonstrated that such a tube would respond to sparks produced at a distance from it.

In 1885 Edison took out a patent on a system of radio communication between ships. However, the patent was not based on the transmission and reception of electromagnetic waves. He later sold the patent to Marconi.

James Clerk Maxwell's theoretical prediction that electromagnetic waves travel at the speed of light was verified in 1888. German physicist Heinrich Hertz made the amazing discovery of radio waves, a type of electromagnetic radiation with wavelengths too long for our eyes to see. He demonstrated the transmission and reception of the electromagnetic waves predicted by Maxwell and thus was the first person to intentionally transmit and receive radio.

Hertz created a transmitting oscillator, which radiated radio waves and detected them using a metal loop with a gap at one side, which he called a resonator. This consisted of a 1m length of thick copper wire, with a small metal circle soldered at each end. The wire was twisted into the shape of a ring with the spheres almost touching each other. When the loop was placed within the transmitter's electromagnetic field, sparks were produced across the gap. Hertz showed in his experiments that these signals possessed all of the properties of electromagnetic waves.

With this oscillator Hertz solved two problems. The first was timing Maxwell's waves. He had physically demonstrated what Maxwell had only theorised: that the velocity of radio waves was equal to the velocity of light. This demonstrated that radio waves were a form of light. Second, Hertz found out how to make the electric and magnetic fields detach themselves from wires and go free as Maxwell's waves. These waves became known as 'Hertzian Waves' and Hertz managed to detect them across the length of his laboratory. This simple resonator was the world's first wireless receiver. Famously, Hertz couldn't see any practical purpose for his discovery. However, his detection led to an increase of experimentation with this new form of electromagnetic radiation.

Nikola Tesla, a Serbian-American inventor, began his research into radio in 1891. Two years later he gave a detailed description of the principles of 'wireless' radio communication to the Franklin Institute in Philadelphia. Tesla's contribution involved refining and improving his predecessor's work. A most important innovation was the introduction of the coupled tuned circuit into his preliminary transmitter design. This was the 'Tesla coil', with its primary and secondary circuits both synchronised to vibrate together in harmony.

Tesla's apparatus contained all the elements that were integrated into radio systems before the early vacuum tube – known then as an oscillation valve – was developed. He first used sensitive electromagnetic receivers, which were different to the less responsive coherers later used by Marconi. Tesla's modifications meant his transmitter could have signalled across the Atlantic, had he thought of such an enterprise. Supplementary work resulted in the development of wireless receivers that also included two synchronised circuits.

After 1890 Tesla experimented with transmitting power by inductive and capacitive coupling using high AC voltages generated with his Tesla coil. He endeavoured to develop a wireless lighting system based on near-field inductive and capacitive coupling. He conducted a series of public demonstrations where he lit incandescent light bulbs from across a stage. In 1893 at the National Electric Light Association, Tesla told his audience that he was certain a system like his could eventually convey 'intelligible signals or perhaps even power to any distance without the use of wires' by conducting it through the Earth.

Tesla would spend most of the decade working on variations of this new form of lighting with the help of various investors. Despite this, none of the enterprises succeeded commercially. Afterwards, the principle of radio communication was publicised widely from Tesla's experiments and demonstrations. Various scientists, inventors, and experimenters began to investigate wireless methods.

Claims have been made that Nathan Beverly Stubblefield, an eccentric farmer from Murray, Kentucky, developed radio between 1885 and 1892, before either Tesla or Marconi. He received widespread attention in early 1902 when he gave a series of public demonstrations of a battery-operated wireless telephone, which could be transported to different locations and used on mobile platforms such as boats.

Stubblefield was convinced other people were stealing his ideas but his devices seemed to have worked by induction transmission rather than radio transmission. Nonetheless, Stubblefield may have been the first to simultaneously transmit audio wirelessly to several receivers, albeit over moderately short distances, while envisaging the eventual development of broadcasting on a national scale.

Stubblefield later became a recluse and lived in a rudimentary shelter near Almo, Kentucky. He died around 28 March 1928 and his body was not discovered until a few days later, having been 'gnawed by rats'. While many later reports state that the cause of death was starvation, at the time of his death a coroner was quoted as saying 'he apparently was a victim of heart disease'. The citizens of Murray, Kentucky, were highly affectionate towards their 'mad radio genius', calling him 'The Father of Radio' and even erecting a monument to him in the town in 1930.

Between 1893 and 1894 a Brazilian priest and scientist, Roberto Landell de Moura, who was commonly known as Roberto Landell, conducted experiments in wireless transmissions. He didn't publicise his achievement until 1900, when he held a public demonstration of a wireless transmission of voice in São Paulo on 3 June. He was granted a Brazilian patent in 1901 before securing three more for a Wave Transmitter, a Wireless Telephone and Wireless Telegraph.

A lack of technical details makes it uncertain which sending technology was being utilised, but if radio signals were employed, then these would be the earliest reported audio transmissions by radio. Although Landell secured patents in Brazil and the United States during the early 1900s, he was unable to acquire enough financial support to further develop his devices.

The Indian physicist Jagadish Chandra Bose publicly demonstrated the use of radio waves in Calcutta in November 1894. Bose set fire to gunpowder and activated a bell using electromagnetic waves and therefore was the first to send and receive radio waves over a significant distance. Bose progressed swiftly with remote wireless signalling and was the first to use semiconductor junctions to detect radio signals. However, he wasn't interested in patenting his work and allowed others to further develop his research.

Oliver Lodge transmitted radio signals on 14 August 1894 at a meeting of the British Association for the Advancement of Science at Oxford University. This was one year after Tesla, five years after Heinrich Hertz and one year before Marconi. On 19 August 1894 Lodge demonstrated the reception of Morse code signalling via radio waves using a coherer. He had upgraded Edouard Branly's coherer radio wave detector by adding a trembler, which displaced clustered particles, thus restoring the device's sensitivity. Lodge had initiated a new system of communication by means of electrical waves that became known as wireless telegraphy.

In August 1898 Oliver Lodge patented 'Electric Telegraphy', which made wireless signals using Induction or Tesla coils for the transmitter and a Branly coherer for the detector. By making the antenna coil or inductance variable, Lodge had made it possible to tune in and select a desired frequency.

In 1895 the physicist Alexander Stepanovich Popov developed a practical communication system based on the coherer. His invention was capable of detecting electromagnetic waves that indicated the presence of electrical discharges, specifically lightning, in the atmosphere. The design of Popov's lightning detector was similar to that of Marconi's wireless telegraph, but Popov's invention focused on receiving rather than transmitting signals. He didn't apply for a patent for this invention.

Popov had expanded upon the work of earlier physicists, such as Heinrich Hertz and Oliver Lodge, but he was the first to incorporate an antenna. Another significant discovery of Popov's came in 1897, when he found that metallic objects could interfere with the transmission of radio waves, a phenomenon known as wave reflection.

On 7 May 1895 Popov demonstrated the transmission and reception of radio waves used for communication at the Russian Physical and Chemical Society. Around March 1896 he reportedly demonstrated the transmission of radio waves between different buildings to the Saint Petersburg Physical Society. This would have been before the public demonstration of the Marconi system. However, other accounts state that Popov achieved these results in December 1897, after publication of Marconi's patent. He later experimented with ship-to-shore communication. Popov died in 1905 and the Russian government didn't press his claim until 1945.

In 1895 the New Zealander Ernest Rutherford arrived in England. The First Baron Rutherford of Nelson was a keen innovator and inventor.

He began using wireless waves as a method of signalling. Sir Robert Ball, who had been scientific adviser to the body maintaining lighthouses on the Irish coast, championed Rutherford's work and hoped he would be able to solve the onerous problem of a ship's inability to detect a lighthouse in fog. Rutherford increased the sensitivity of his apparatus until he could detect electromagnetic waves over a distance of several hundred metres.

Karl Ferdinand Braun made two major contributions to the development of radio. The first was the introduction of a closed tuned circuit in the generating part of the transmitter, and its separation from the antenna by means of inductive coupling. Around 1898 he invented a crystal diode rectifier or 'Cat's whisker diode'. Braun's invention bridged a much longer distance.

Our chronological path now leads us to the person who is widely recognised as the true father of radio. Guglielmo Marconi was fascinated by Heinrich Hertz's discovery of radio waves, and realised that if they could be transmitted and detected over long distances, wireless telegraphy could be developed for commercial purposes. He started experimenting in 1894 and installed rough aerials on opposite sides of his family's garden in Bologna, Italy. His aerials were tin plates mounted on posts. Marconi managed to receive signals over a distance of 100m, and by the end of 1895 had extended the distance to over a mile. Marconi offered his telegraph system to the Italian government, but they turned it down.

The British Post Office was more receptive and Marconi moved to London. In February 1896 he constructed his transmitter on the roof of the Central Telegraph Office, and a receiver on the roof of a building called 'GPO South' in Carter Lane, 270m away. His later transmissions were detected 2km away, and on 2 September at Salisbury Plain the range was increased to 12km.

Marconi received the first wireless patent from the British government. In part, it was based on the theory that the communication range increases substantially as the height of the aerial above ground level increases. On 12 December 1896 Marconi gave his first public demonstration of radio at Toynbee Hall, London.

In 1897 Marconi established the Wireless Telegraph and Signal Company at Chelmsford and the world's first radio factory was opened there, employing fifty people. On 11 May 1897 tests were carried out to establish that contacts were possible over water. A transmitter was constructed at Lavernock Point, near Penarth, and the transmissions were received on the other side of the Bristol Channel, at the Island of Holm, a distance of 6km.

In November 1897 the first permanent radio installation, 'Needles Hotel Wireless Station', was installed at Alum Bay, on the Isle of Wight, by the Wireless Telegraph and Signal Company. Alum Bay was an isolated but striking strip of coastline that provided open water straight to the mainland just as far as Marconi's equipment's top range.

Marconi managed to transmit to two hired ferryboats and to another station in Bournemouth. Alum Bay may have helped launch wireless but this didn't impress the inventor's landlord. The Royal Needles Hotel subsequently raised Marconi's rent so he dismantled the station at the end of May 1900 and moved further down the coast.

The *Daily Express* was the first newspaper to obtain news by wireless telegraphy in August 1898. In December 1898 Marconi installed radio equipment on the Royal Yacht *Osborne*, which was moored at Cowes. Regular messages were relayed from the yacht and from Osborne House, also on the Isle of Wight. The messages were then passed on to Buckingham Palace. The Queen received 150 bulletins on the Prince of Wales' health from the yacht, where he was convalescing. The Prince operated the equipment on the Royal Yacht while Marconi operated the equipment in Osborne House. Around the same time wireless communication was established between the *East Goodwin* Lightship and the South Foreland Lighthouse.

In 1899 Marconi was on board the HMS *Defiant* and observed proceedings as the ship's captain gave orders to three cruisers in controlled manoeuvres via radio for the first time. The first telegraph message was sent across the English Channel on 27 March 1899. It was sent from South Foreland to Wimereux, in France, by Marconi. The success of the demonstration resulted in lighthouses throughout the UK being fitted with wireless sets.

On 17 March 1899 Marconi garnered a lot of publicity when wireless telegraphy was used to save a ship in distress in the North Sea. The three-masted ship *Elbe* was sailing to Hamburg with a cargo of slates. A thick fog was prevailing at the time when the ship went ashore on the Goodwin Sands. The *East Goodwin* Lightship heard the signals and communicated by wireless telegraphy to the South Foreland Lighthouse.

From there telegraphic messages were sent to the authorities, and lifeboats at Ramsgate, Deal and Kingsdown were put on standby. Fortunately, the lifeboats weren't required as the *Elbe* was able to refloat eight hours later. Nevertheless, this was the first occasion in which lifeboats had been alerted by the means of wireless.

About this time Marconi began to develop tuned circuits for wireless transmission so that a wireless could be tuned to a particular frequency. He patented this on 26 April 1900, under the name of 'Tuned Syntonic Telegraphy'. His next project was to send a signal across the Atlantic. He convinced investors to spend £50,000 on the transatlantic project and purchased land in Poldhu, Cornwall. This site was chosen by Marconi because it stood directly opposite Cape Cod, where its sister radio station was to be built. The site was also chosen for its remoteness, to keep the project out of the public eye and out of the newspapers.

It was a massive undertaking that dwarfed anything he had built before. Construction work began in October 1900 when around 400 wires were suspended in an inverted cone shape from twenty masts. Infuriatingly, the system was blown down during a storm, so a temporary aerial was hastily assembled, using two surviving masts, to let the transatlantic experiments carry on. A year later the Poldhu Wireless Station had successfully transmitted signals to ships at distances over 321km. Nonetheless, the transatlantic project remained Marconi's main goal.

On the other side of the Atlantic, the Cape Cod site was eventually abandoned. Numerous difficulties, including severe weather, necessitated the move of the receiving station from Cape Cod to St John's Newfoundland, which was also 965km closer to Cornwall.

Marconi travelled across the Atlantic to supervise proceedings from that end. Due to time and financial constraints, he opted not to build a masted receiving antenna array. The original receiving antenna in Newfoundland was 10cm in diameter and was held aloft by a balloon, which was ripped apart in a storm. The first attempt to send signals across was made in November 1901, but failed when one of two balloons holding the aerial wire aloft broke its mooring and floated away.

At 12.30am on 12 December 1901, at Signal Hill in St John's, Newfoundland, Marconi heard three faint clicks through the earphones of his wireless receiver – the Morse code letter 'S' – and a new era was born. The receiving aerial, 180m of wire, was held aloft by six kites flying at an altitude of 120m.

The British government and Admiralty were greatly impressed and many people wanted to invest in the new technology. Demand grew and large numbers of ships carried the new apparatus, which saved many lives at sea. One of the most famous occasions was when the RMS *Titanic* sank. There were two wireless operators on board the ship that fateful evening of 14 April 1912. Jack Phillips was the senior operator on board the doomed vessel along with Harold Bride, the junior telegraphist.

The wireless room had been kept busy with commercial traffic since departing from Southampton. The transmission equipment had developed a fault as a result and in turn this had led to a decrease in power output. According to the Marconi manual and company policy, telegraphists were not to attempt to fix this particular component but wait until getting into port whereby a Marconi engineer could be called to repair the fault. Instead, Phillips made the decision to troubleshoot and fix the problem, working through his allotted time to sleep. If he had taken the decision to wait, it's likely that they would not have had sufficient power to contact the rescue ships later.

Phillips eventually repaired the fault in the afternoon and had been working hard to clear the backlog of messages, sending them via Cape Race in Newfoundland. There had been several communications over the wire from various ships warning that the *Titanic* was heading towards an ice field. Most of these had not been conveyed to the bridge for various reasons, but similar warning messages had been delivered to the captain earlier that day and a lookout had been posted.

Phillips has since been heavily criticised for having told the radio operator of the RMS *Californian*, 'Shut up! I am busy, I am working Cape Race,' when interrupted on-air by his counterpart. The use of the words 'shut up' was a common short form among wireless operators to politely ask other operators to 'keep the line free' and had absolutely nothing to do with any conceit.

The *Titanic* struck an iceberg at 11.40pm that night and began to sink. Captain Edward Smith entered the wireless room and told the men to prepare to send out a distress signal. Shortly after midnight the captain came in again and told them to send out the call for assistance and gave them *Titanic*'s estimated position. Both men elected to stay at their positions for as long as possible to help communicate with the ships coming to assist with the rescue.

Phillips began sending out the distress signal code CQD. Bride had to remind him that the new call was SOS. Phillips stayed at the wireless set frantically tapping away at the key. While all this was going on, Bride relayed messages to Captain Smith about which ships were coming to the vessel's assistance. Following one visit to the bridge, the telegraphers were warned that the forward part of the ship was flooded and advised to put on more clothes and grab lifesaving equipment.

The power was cut shortly after 2am, rendering the wireless equipment useless. Captain Smith arrived and told the men they were relieved of their duty and to 'shift for themselves'. As the men made their preparations to leave, another crew member attempted to steal Phillips' lifebelt. There followed a brief scuffle which ended when Phillips knocked the crew member out. By this time the water was beginning to flood the wireless room and they both rushed out, leaving the unconscious crewman where he fell. The men then split up, with Bride heading forward and Phillips heading aft. This was the last time Bride saw Phillips alive.

Harold Bride was washed overboard as the *Titanic*'s boat deck flooded. Nevertheless, he managed to scramble onto the upturned lifeboat Collapsible B and was rescued by the RMS *Carpathia* later in the morning. Despite being injured, he helped the *Carpathia*'s wireless operator transmit survivor lists and personal messages from the ship.

The fate of Phillips is less certain. There are several uncorroborated reports about what happened to him. One account stated that he had made it to the same lifeboat that Bride had reached but subsequently died of exposure. Another report had him clinging to the side of one lifeboat but he was too weak to clamber aboard. None of these accounts were ever satisfactorily confirmed and unfortunately his body was never recovered.

The *Titanic* took less than three hours to sink, taking almost 1,500 passengers and crew with her. Phillips had only celebrated his twenty-fifth birthday on the ship two days before the disaster.

Titanic's wireless set had a nominal working range of 463km, but signalling more distant stations was possible. At night, ranges of up to 3,200km were attained with sets of similar design. The 'T' type aerial that was used offered greater power and sensitivity, both fore and aft, therefore optimised performance could be expected when the ship was pointed either towards or away from a distant station. The ability to send signals over great distances helped to summon assistance much quicker and undoubtedly saved many lives. Without Marconi's technology, all those aboard would have perished and the fate of the ship would have remained a mystery.

In 1909 Marconi shared the Nobel Prize in Physics with the German physicist Karl Ferdinand Braun, the inventor of the cathode ray tube. In his acceptance speech he freely admitted he didn't really understand how his invention worked. He always regarded himself as much more a tinkering engineer than a scientist.

At the outbreak of the First World War all Marconi's energy was diverted to the war effort. German technology was, in many ways, superior to that of the Allies. Nevertheless, when it came to radio, Marconi's expertise helped Britain immensely. British wireless operators had discovered how to intercept German radio signals sent from the trenches and many surprise attacks were foiled due to the advance warnings given by Marconi's equipment.

The Marconi technology also proved particularly useful at sea. During the Battle of Jutland, the primary and most critical naval engagement of the war, the substantial battle fleet of the German Imperial Navy had the potential to inflict mayhem on British maritime lifelines. To counteract this, the Royal Navy patrolled the mouth of the Skagerrak strait, bottling the Germans into the Baltic Sea.

On 30 May 1916 the German fleet attempted a break-out. Thanks to radio intercepts, Lord Jellicoe, the British admiral, had advance warning of the move, giving him time to order his ships to battle stations. The subsequent battle turned out to be inconclusive. Under cover of night, the German fleet slipped back to the security of the Baltic ports, where it remained until the end of the war. A year later the entire imperial battle fleet was escorted to Scapa Flow and scuttled.

With Europe at war, and wireless engineers immersed in the war effort, it was left to the Americans to advance long-distance wireless telephony. In October 1915 the American Telephone & Telegraph Company, working with the Western Electric Company, successfully broadcast speech and music from Arlington, Virginia, to the Eiffel Tower in Paris, a distance of some 5650km. For this, the first successful transatlantic telephonic relay, as many as 500 valves were required for the transmitter. An even greater distance, 8,000km, was achieved a year later when speech was again relayed from Arlington, this time to Honolulu.

When the Radio Act of 1912 was legislated under US Federal law, licensing fell under the auspices of the Department of Commerce. There is no comprehensive record of the stations licensed under this act. The department had no authority to deny a licence to anyone who requested one, and didn't regulate frequencies or power.

The aspirations of the early radio pioneers didn't include the broadcasting of music and information into homes using wireless. All the same, some people saw it as a serious wireless alternative to the Bell telephone.

Conveying voice or music by radio required a continuous-wave transmitter. In 1902 Danish engineer Valdemar Poulsen invented an arc converter as a generator of continuous-wave radio signals. Beginning in 1904, Poulsen used the arc for experimental radio communication from Lyngby to various sites in Denmark and Great Britain.

The arc was formed between copper and carbon electrodes enclosed in a gas-tight vessel containing either coal gas or hydrogen, and inserted directly in the aerial-earth circuit. The arc transmitter emitted a fixed constant note in the receiver's headphones when the Morse key was pressed. This enabled signals to be sent over very much longer ranges than by spark transmission. By connecting a microphone in the direct current supply circuit of the arc, it was seen that this could vary the current flowing through the arc and so be used for the transmission of wireless telephony. The high-frequency continuous wave produced by the arc acted as a carrier wave for the small varying waves of low-frequency speech that were superimposed upon it. This 'modulated' wave was then picked up in the aerial of the receiver, and the low-frequency current was detected and separated from the carrier before being received through the headphones.

The radio-telephone years of 1900-1920 were known more for the rival voice transmission technologies than for broadcasting. While spark was quickly rejected as too noisy and the alternator as too costly, it was the many versions of the Poulsen arc that clearly dominated radio-telephone inventions and early broadcasting for an audience.

Surprisingly, Marconi saw no need for voice transmission. He felt that the Morse code was adequate for communication between ships and across oceans. Marconi didn't anticipate the development of the radio and broadcasting industry and he left the early experimentation with wireless telephony to others.

Professor Reginald Aubrey Fessenden's technology and circuit arrangements were very different to Marconi's. He tried all the various methods of generating wireless signals in the early days: by spark, by arc and by the high-frequency alternator. His work was dominated by his interest in transmitting words without wires. Fessenden's equipment included a spark transmitter, using a Wehnelt interrupter working a Ruhmkorff induction coil. In 1899 he noted that when the key was held down for a long dash, the odd wailing sound of the Wehnelt interrupter could be clearly heard in the receiving telephone. This suggested to him that by using a spark rate far above voice band, wireless telephony could be achieved.

On 23 December 1899 Fessenden succeeded in transmitting speech, albeit highly distorted, over a distance of 1.5km. By 1904 fairly satisfactory speech had been transmitted by the arc method. Nevertheless, Fessenden remained an impassioned supporter of the continuous wave method of transmission.

He developed his new high-frequency alternator-transmitter, showing its utility for point-to-point wireless telephony, including interconnecting his stations to the wire telephone network. Fessenden placed a carbon microphone directly in line between his alternator and the antenna lead.

Fessenden also invented the heterodyne effect. In this, a received radio wave is combined with a wave of a frequency slightly different from the carrier wave. The intermediate frequency wave that is produced as a result is easier to amplify, and can then be demodulated to generate the original sound wave.

Marconi's transatlantic experiments had captured the public's imagination, but Fessenden had also been conducting his own transatlantic transmission experiments from the National Electric Signalling Company at Brant Rock in Massachusetts. To carry out these experiments Fessenden's Company built a station at Machrihanish in Scotland, installing exactly the same equipment as at Brant Rock. After numerous attempts it became evident that no signals were coming through from Scotland. Fessenden sent one of his best engineers, James Armor, to investigate and, in January 1906, Armor sent a telegram saying that Machrihanish was receiving the signals from Massachusetts loud and clear.

Encouraged by this achievement, Fessenden enhanced the effectiveness of his high-frequency alternator, and with a new type of umbrella antenna of his own design, both stations were in regular communication. In June a small testing station was built at Plymouth, 17km from Brant Rock. The engineers used voice transmission to communicate with each other.

In November a letter was received from James Armor containing the astounding news that he had clearly heard the complete conversation of Adam Stein, Fessenden's chief engineer, at Brant Rock, telling the operator at Plymouth 'how to run the dynamo'. The first human voice to be transmitted across the Atlantic, therefore, was that of Adam Stein.

The Machrihanish tower collapsed during a severe winter storm on 5 December 1906. It was never rebuilt and so Fessenden's transatlantic trials came to a rapid conclusion. Instead, he decided to concentrate on developing voice transmission. On Christmas Eve 1906, from his workshop in Chestnut Hill, Massachusetts, Fessenden sent a Morse message alerting all ships at sea to expect an important transmission. What they heard that night was the first public broadcast of the human voice.

Fessenden stepped up to the asbestos-covered microphone and proceeded to give a brief description of the forthcoming broadcast. He then played an Edison wax-cylinder recording of Handel's *Largo*. Fessenden then treated his listeners to his rendition of *Oh Holy Night* on the violin, actually singing the last verse as he played.

Adam Stein was due to make an announcement but panicked and was unable to say a word. Consequently, he became the first person to suffer from 'mic fright'. Fessenden's wife, Helen, and his secretary, Miss Bent, had promised to read passages from the Bible, but when the time came to perform they also froze and Fessenden took over. He concluded the broadcast by extending Christmas greetings to his listeners and asked them to write and report to him on the broadcast wherever they were. The broadcast was successfully repeated on New Year's Eve.

Fessenden's claim that these two broadcasts were the first to broadcast speech and music is disputed. There are claims that the first broadcast of music actually took place on 15 June 1904 at the university in Graz, Austria. Professor Otto Nussbaumer personally yodelled an Austrian folk song into a microphone and the transmission was heard at a distance of 22m. Then there is the claim that in the spring of 1906, a full six months before Fessenden's broadcast, a wireless operator on board the USS *Missouri* played the melody of the folk song *Home Sweet Home* by varying the speed of the spark generator. Nevertheless, Fessenden can rightly claim that his was the first intentional broadcast.

Fessenden's methods were extremely primitive when compared to today's standards. They were, nevertheless, the first real departure from Marconi's damped-wave-coherer system for telegraphy, which other experimenters were merely emulating or adapting. They were the first ground-breaking steps towards radio communications and radio broadcasting.

Following the *Titanic* sinking after colliding with an iceberg in the Atlantic, Fessenden announced that he had 'bounced signals off icebergs by radio, measuring the distance.' His invention could realistically be described as the forerunner of radar.

Reginald Aubrey Fessenden died in Bermuda on 22 July 1932. His tomb was inscribed with these words: 'By his genius, distant lands converse and men sail unafraid upon the deep.'

It's unclear exactly who was the first person to conduct a telephony broadcast in Britain but most historians point to Lieutenant Quentin Crawford. In 1907 the British Admiralty authorised him to create an experimental radio station on board HMS *Andromeda*. Using the call sign QFP, he adapted the spark wireless transmitter on board to broadcast a programme featuring music and speech for the benefit of the Royal Navy fleet in Chatham dockyard. Crawford's historic inaugural broadcast was a patriotic concert programme performed by navy personnel. This transmission was widely perceived to be a success but the navy decreed that the broadcast be kept secret.

There are also several reports that a British amateur had broadcast wireless telephony that year. H. Anthony Hankey had used a portable Poulsen Arc transmitter located in Aldershot in a demonstration to members of the military in Midhurst, Kent. Hankey's broadcast consisted of a few songs and monologues performed personally by him. Many people believe this may have been the first wireless telephony broadcast in the UK. Nevertheless, unlike the Admiralty, the army was unenthusiastic about the adoption of wireless in the years before the First World War and took it no further.

Lieutenant Crawford's naval broadcast from Chatham is claimed as the first radio music broadcast from a ship. A year later, army radio operators at Sandy Hook in New Jersey made several experimental broadcasts of music for reception at Fort Wood on Bedloe's Island, the location of the famous Statue of Liberty.

Harry Grindell Matthews was a British wireless experimenter whose first telephony broadcasts featured him whistling and playing the banjo.

He demonstrated the versatility of wireless telephony by being the first to communicate with a moving aircraft. On Saturday, 23 September 1911, at Ely Racecourse in Cardiff, Matthews established radio contact with pioneer aviator Bentfield Charles Hucks, who was flying at a height of 215m and at a speed of 96km per hour.

Matthews called his invention an Aerophone device, which soon attracted the attention of the government. He was asked to demonstrate its capabilities to the Admiralty. Matthews agreed, but requested that no experts be present at the scene. When four of the observers dismantled part of the device before the demonstration began and took notes, Matthews stopped the demonstration and sent observers away. Newspaper reports backed Matthews' side of the story but The War Office denied any meddling. They initially claimed that the demonstration was a failure but later stated that the affair was just a misunderstanding.

In 1914, after the outbreak of the First World War, the British government announced an award of £25,000 to anybody who could remotely control unmanned vehicles. Matthews claimed that he had produced a remote control system that used selenium cells. He demonstrated it with a remotely controlled boat to representatives of the Admiralty at Richmond Park's Penn Pond. It's reported that he received his payment but the Admiralty never used the invention.

Matthews claimed to have perfected a way of transmitting energy without wires. He claimed that his 'Death Ray' could stop the ignition system of a spark-ignition engine and, with enough power, would be able to shoot down aeroplanes and explode ammunition dumps.

The Air Ministry was wary because of previous bad experiences with would-be inventors. Nevertheless, Matthews was requested to demonstrate his ray on 26 April 1924 to the armed forces. In Matthews' laboratory they saw how his invention switched on a light bulb and cut off a motor. Unfortunately, he failed to convince the officials, who suspected it might be a confidence trick.

When the British Admiralty requested a further demonstration, Matthews refused to give it and announced that he had an offer from France. Swiftly, the High Court in London granted an injunction that forbade Matthews from selling the rights to the 'Death Ray'. When Major Wimperis arrived at his laboratory to discuss a new agreement, Matthews had already flown to Paris.

Questions were asked in parliament what the government intended to do to stop Matthews selling his 'Death Ray' to a foreign power. The Under Secretary for Air answered that Matthews was not willing to let them investigate the ray to their satisfaction. The government said it was vital that Matthews demonstrated the ray's ability to stop a petrol motorcycle engine in the conditions that would satisfy the Air Ministry. In return, he would receive £1,000 and further consideration. From France, Matthews answered that he was not willing to give any further proof.

Matthews then tried to market his invention to America. He was offered $25,000 to demonstrate his beam to the Radio World Fair at Madison Square Garden. He refused again and claimed, without substance, that he wasn't allowed to demonstrate it outside Britain.

He later claimed to have invented a device that projected pictures onto clouds, aerial mines and a system for detecting submarines. However, his reputation preceded him and the government was no longer interested in his concepts.

In 1904 Sir John Ambrose Fleming invented the thermionic diode or valve, which assisted the detection of high-frequency radio waves. He'd adapted Edison's electric lamp and added a second element called a plate. The valve contained a carbon filament that was made incandescent by an electric current. The fibre was sealed in the glass bulb and all of the air was removed. Around the filament, but not touching it, was a metal pipe attached to a wire and sealed through the bulb, while the terminals of the filament and the cylinder were fixed to a base. The Fleming Diode acted as a detector and rectifier of the incoming high-frequency alternating currents picked up by the receiver's aerial, and transformed these into direct current to which the headphones responded.

John Ambrose Fleming's invention was a major step forward in wireless technology as it was considerably more efficient as a radio wave detector than coherers or magnetic detectors. Valves became fundamental components in radios, as well as television sets and computers for fifty years, before being superseded by the transistor.

American Lee de Forest was convinced there was a great future in wireless telegraphy but Marconi was already making impressive progress in both Europe and the United States. One drawback to Marconi's approach was his use of a coherer as a receiver. This method was slow, insensitive and unreliable. To provide a permanent record of the transmission, it had to be tapped to restore operation. He was determined to devise a superior system, including a self-restoring detector that could receive transmissions by ear, thus giving it the ability to receive weaker signals and also enabling faster Morse code sending speeds.

After a succession of short-lived jobs with various communication companies, de Forest decided to strike out on his own. In the autumn of 1901 Marconi had been hired by the Association Press to provide reports for the International Yacht Races in New York, and de Forest struck a deal to provide a similar service for the smaller Publishers' Press Association.

The race deal turned out to be a disaster for de Forest as his transmitter broke down. In a fit of rage, he threw it overboard and had to revert to using a spark coil as an alternative. Unfortunately, none of the other companies in attendance had effective tuning for their transmitters, so only one could transmit at a time without causing reciprocal interference. Although an attempt was made to have the different systems avoid disputes by rotating operations over five-minute intervals, the arrangement broke down quickly, resulting in signals being lost in electronic mush. De Forest mournfully conceded that under the circumstances, visual semaphore 'wig-wag' flags were the only successful 'wireless' communication during that period.

In January 1907 de Forest patented the triode valve, which he called the 'audion'. He modified Ambrose Fleming's design by adding a grid to control and amplify radio and sound waves. Not only was the audion a very efficient detector of wireless waves, it also acted as an amplifier. This made it possible for the reception of transmissions over far greater distances than had been possible before. This was one of the most important strides in the development of communication as many of the previous problems had now been overcome.

Later that year he formed the de Forest Radio Telephone Company and began seeking investors. His first experimental broadcasts were transmitted from the top floor of the Parker Building on Fourth Avenue and Nineteenth Street in New York in February 1907. Although the broadcast consisted mainly of Columbia gramophone records, it was widely considered successful. At times it was the victim of quite severe interference from Morse code transmissions and some noticeable signal fading due to atmospheric conditions.

As de Forest (and hundreds of other inventors) unfortunately discovered, you had to be a promoter as well as an inventor.

Arguably, there was no wireless and radio inventor who was more associated with controversy than Lee de Forest. He was occasionally linked with unsavoury business promoters and was accused frequently of dishonest business practices. Opinions are divided about him – he seems to have been vilified or sanctified in equal measure – yet the evidence robustly indicates that he was the first to want to use the wireless for more than two-way commercial message traffic. De Forest fought for decades to persuade the scientific community that he deserved to be known as the 'Father of Radio'. He spent millions in court battles trying to validate and revalidate his patents.

In 1909 de Forest invited his eminent mother-in-law, Harriet Stanton Black, to give the world's first broadcast talk regarding the women's suffrage movement. De Forest was a music lover and he soon came up with the idea of having a live performer on the radio. In September 1907 Eugenia Farrar, the Swedish soprano, sang a few songs from the USS *Connecticut* in the Brooklyn Navy Yard. In an early company article, de Forest predicted, 'It will soon be possible to distribute grand opera music from transmitters placed on the stage of the Metropolitan Opera House by a Radio Telephone station on the roof to almost any dwelling in Greater New York and vicinity.'

That prediction eventually became reality on 12 January 1910 when he conducted an experimental broadcast of the live performance of the opera *Tosca*. The next day he broadcast a performance with the participation of the Italian tenor Enrico Caruso. He installed a 500-watt transmitter in an attic room of the opera house for these broadcasts and the aerial was strung between two bamboo fishing rods on the roof. To capture the performances, he placed one microphone on the stage and another in the wings. The audiences for these broadcasts were largely journalists who were invited to experience the events. They were huddled around receivers placed strategically in different parts of New York.

By the start of 1916 de Forest had refined his audion to be used as an oscillator for the radio-telephone. He sold it to the telephone company as an amplifier of transcontinental phone calls. Later that year he broadcast the first radio advertisements from experimental radio station 2XG in New York City. Incidentally, the advertisements were for his company's products. He went on to sponsor radio broadcasts of music, but received little financial backing.

De Forest transmitted the first presidential election report by radio in November 1916, from his station in The Bronx, assisted by a wire supplied by the Republican newspaper *The New York American.* He sent bulletins out every hour, and between the bulletins listeners heard *The Star-Spangled Banner* and other anthems, songs and hymns. The broadcast lasted approximately six hours until signing off about 11pm. However, it ended before Woodrow Wilson came from behind to win, so many of the listeners heard the wrong candidate declared the winner.

A few months later, de Forest moved his transmitter to High Bridge, New York. He had been granted a licence from the Department of Commerce for an experimental radio station, but had to cease all broadcasting when the US entered the First World War in April 1917.

De Forest resumed his broadcasts from High Bridge after the war, but came into conflict with the US Federal Inspector, so moved his operations to California in April 1920. He operated his station 6XC from the California Theatre until November 1921, after which the transmitter was moved to Oakland and 6XC became KZY, the Rock Ridge Station.

Lee de Forest moved on to work on a variety of non-radio technical devices, most notably his Phonofilm system, a process to make the movies talk by adding a synchronised optical soundtrack to the film. In his final years, he was disillusioned at what radio programming had become. Speaking to reporters, he asked, 'Why should anyone want to buy a radio? Nine-tenths of what one can hear is the continual drivel of second-rate jazz, sickening crooning by degenerate sax players, interrupted by blatant sales talks.'

John Stone Stone was an American physicist and inventor. He first worked for the research and development department of the Bell Telephone Company. However, he is probably better known for his influential work developing early radio technology, in particular for improvements in tuning. He recognised that his earlier work on resonant circuits on telephone lines could be applied to improve radio transmitter and receiver designs. Stone used his knowledge of electrical tuning to develop a 'high selectivity' approach to reduce the amount of interference caused by static and signals from other stations.

In 1902 he formed his own company in Boston. Beginning in 1905, Stone demonstrated radio-telegraphy stations to the US Navy using spark transmitters and electrolytic detectors and by the end of 1906, the government had purchased five ship and three land installations.

The company's first commercial radio-telegraph link was between the Isle of Shoals and Portsmouth, New Hampshire, which operated during the summer of 1905, replacing a failed Western Union telegraph cable. Despite his advanced designs, the Stone Telegraph and Telephone Company failed in 1908. Its assets, including its valuable portfolio of patents, were sold to Lee de Forest's company and Stone spent the remainder of his career as an engineering consultant.

Before the First World War, receivers were mainly crystal sets, which were exceedingly insensitive and non-selective. They were connected to a pair of headphones and required a long aerial. In Britain, the new technology was strictly controlled by the Post Office. It was reasonably simple to acquire a receiving licence but a much more complicated proposition to obtain permission to use a transmitter.

The British Post Office had to be satisfied that the applicant had suitable engineering qualifications, or knowledge, to operate the transmitter. Transmitter output power was restricted to 10 watts, and use was only permitted for scientific research or for something of use to the public. Only a small number of radio amateurs were transmitting before the First World War.

As in many other wars, the First World War hastened the development of technology that was useful for the war effort. Although valves had been produced since 1904, the inability to produce a good vacuum meant that these devices were unreliable and had a short life.

Irving Langmuir was an American physicist, best known to the electrical industry as the inventor of the gas-filled tungsten lamp. He discovered that filling the bulb with an inert gas, such as argon, could lengthen the lifetime of the filament. He also discovered that by curling the thread into a tight coil, he could enhance its efficiency. Langmuir developed a method of producing an excellent vacuum, but his first major development was the improvement of the diffusion pump, which ultimately led to the invention of the high-vacuum tube (the vacuum tube is more commonly known as a valve in Britain and Europe).

French military scientists used Langmuir's technique to produce a reliable and efficient triode valve, which was called the 'R' valve. It was used in military communication equipment and was produced in large numbers. After 1916 lamp manufacturers Osram also produced the valve in England.

When the war ended, many of these valves appeared on the surplus market and therefore were readily obtainable. A lot of people were interested in the new technology and began building receivers, and so the number of radio amateurs grew rapidly. The new valves made it possible to easily transmit high-quality speech and music, and allowed high-sensitivity receivers to be developed.

The first regularly scheduled broadcasts had begun in America in 1912. The person responsible for these was Charles David Herrold. He had enrolled at Stanford University in 1895. While there, he was inspired by reports of Marconi's demonstrations and began to experiment with the new technology.

Illness forced Herrold to withdraw from the university without graduating after three years. When fully recovered he moved to San Francisco, where he developed a number of inventions for surgery, dentistry and underwater lighting. Unfortunately, the infamous 1906 San Francisco earthquake struck and Herrold lost everything.

Subsequently, he took an engineering teaching position at Heald's College of Mining and Engineering in Stockton, California. During his spell there, one of his research projects included the remote detonation of mines using radio signals. During his three years at the college, he received further inspiration from the novel 'Looking Backward' by Edward Bellamy, which envisaged the dissemination of entertainment programming over telephone lines to individual homes. Herrold began to speculate about the possibilities of using radio signals to distribute the programming more efficiently.

When he returned to San Francisco in 1909 he opened his own school. The Herrold College of Wireless and Engineering was located in the Garden City Bank Building at 50 West San Fernando Street in San Jose, where a huge 'umbrella-style' antenna was constructed on the roof of the building.

The college's primary purpose was to train radio operators for handling communications aboard ships or staffing shore stations. Although Herrold never held a degree, his students knew him respectfully as 'Doc'. By all accounts he was an excellent teacher. His spare time was spent inventing ways to make his wireless radio inventions talk.

Herrold's primary radio-telephone effort was towards developing a commercial system suitable for point-to-point service. Working with his assistant Ray Newby, he initially used high-frequency spark transmitters. Nevertheless, as the limitations of the high-frequency spark soon became apparent, he diverted to refining versions of the Poulsen arc, which was more stable and produced better sound.

In 1912 Herrold was hired as chief engineer of the National Wireless Telephone and Telegraph Company in San Francisco. They hoped that he could develop a highly profitable point-to-point 'arc fone' radio-telephone. Herrold produced a system with good quality sound, informally described as 'shaving the whiskers off the wireless telephone'. Despite the low power, a number of successful tests for the US Navy were reported.

Unfortunately, the professional partnership between Herrold and NWT&T was not to be a lengthy or happy one. In late 1913 Herrold resigned and then sued NWT&T on the grounds that he had not been fully recompensed for his time and effort. The company counter-claimed that they had honoured the terms of his contract. Moreover, the company ultimately abandoned most of the improvements made by Herrold. The judge sided with NWT&T and rejected Herrold's claim. In addition, despite his attempts to create a transmission system that didn't encroach upon the patents of the Poulsen arc, there was a degree of uncertainty that he had actually achieved this objective.

Charles Herrold didn't claim to be the first to transmit the human voice but he claimed to be the first to conduct 'broadcasting'. He coined the terms 'narrowcasting' and 'broadcasting' respectively, to identify transmissions destined for a single receiver, such as that on board a ship, and those transmissions destined for a general audience. Herrold was the son of a Santa Clara Valley farmer and the term 'broadcasting' was originally a farming idiom, meaning to spread seed over a large expanse.

Herrold made the first planned radio broadcast from his radio school in 1912. The station announced itself as 'San Jose Calling' initially as there were no call letters assigned at the time. He was later awarded two permits: 6XE for portable operations and 6XF, a standard experimental licence.

To help the radio signal to spread in all directions, he designed some omnidirectional antennas, which he mounted on the rooftops of various buildings in San Jose. Herrold also claims to be the first broadcaster to accept advertising. He exchanged publicity for the Wiley B. Allen Company, a local record shop, for records to play on his station. Herrold's wife, Sybil, later recalled that she took part in several of the Wednesday night programmes where she played recordings that had been requested by the listeners.

Herrold's ultimate transmitter design featured a water-cooled microphone linked to six small arcs burning in liquid alcohol.

A review of a Christmas 1916 concert complimented the good audio quality of the 'Herrold-Portal aerial system of telephony', stating, 'It was as sweet and beautiful as if it had been played and sung in the next room.'

Despite the popularity of the broadcasts, they received only scant local attention, and were largely unheard of outside the immediate San Jose area. The broadcasts eventually finished on 6 April 1917, when all civilian station operations were suspended as a result of the United States' entry into the First World War.

Eventually, on 9 December 1921, a licence for San Jose, with the randomly assigned call sign of KQW, was issued to Herrold. Operation of the broadcasting station was financed by sales of radio equipment by the Herrold Radio Laboratory, but by 1925 the costs for KQW had grown considerably and the station was sold to the First Baptist Church of San Jose. Two stipulations of the sale were that Herrold be kept on as the director of programmes and, secondly, the station's sign-ons had to include the statement: 'This is KQW, pioneer broadcasting station of the world, founded by Dr Charles D. Herrold in San Jose in 1909.'

Herrold ended his association with KQW in 1926 and began working as a salesman for KTAB in Oakland, California. He later became a repair technician and a janitor, and died in a Californian retirement home, aged seventy-two. Towards the end of his life he sought recognition for his pioneering broadcasts. The general consensus is that he was the first to transmit regular entertainment broadcasts, therefore giving him a legitimate claim to the self-proclaimed title of 'Father of Broadcasting'.

American engineer and inventor Edwin Howard Armstrong invented three of the basic electronic circuits underlying all modern radio, radar and television. In 1914, while still an undergraduate, he patented the regenerative circuit, which produced amplification hundreds of times greater than previously attained. This amplified signals to an extent that receivers could use loudspeakers instead of headphones.

While serving as a major in the US Army Signal Corps during the First World War, he developed the super-heterodyne receiver. This circuit made radio receivers more sensitive and selective and is still extensively used today.

Other people ultimately claimed many of Armstrong's inventions in patent lawsuits. In particular, the regenerative circuit, which he patented in 1914 as a 'wireless receiving system,' was patented by Lee de Forest two years later. De Forest then sold the rights of his patent to AT&T.

Armstrong was incredibly fond of grand stunts. In 1923 he climbed the WJZ aerial array situated at the top of a twenty-storey building in New York City and reportedly performed a handstand. He arranged to have photographs taken and delivered to Marion MacInnis, secretary to RCA President, David Sarnoff. Armstrong and MacInnis married later that year. A publicity photograph was made of him presenting Marion with the world's first portable super-heterodyne radio as a wedding gift.

Between 1922 and 1934 Armstrong found himself embroiled in a patent war. On one side were Armstrong, RCA, and Westinghouse, and de Forest and AT&T on the other. This action was the longest patent lawsuit ever litigated up to that period. Armstrong won the first round, lost the second, and the third ended in stalemate. Ultimately de Forest was granted the regeneration patent at the Supreme Court, though many people today believe this result was due to a misunderstanding of the technical facts.

The legal battles had one serendipitous outcome for Armstrong. While he was preparing equipment to oppose a claim made by a patent attorney, he 'accidentally ran into the phenomenon of super-regeneration', where, by rapidly 'quenching' the valve oscillations, he was able to achieve even greater levels of amplification. In 1922 Armstrong sold his super-regeneration patent to RCA. This eventually made him RCA's largest shareholder, and he noted that 'The sale of that invention was to net me more than the sale of the regenerative circuit and the super-heterodyne combined.'

While the lawsuits dragged on, Armstrong was already working on another significant invention. He created wide-band frequency modulation radio or FM. Rather than varying the amplitude of a radio wave to create sound, Armstrong's method varied the frequency of the wave instead. FM radio broadcasts delivered a much clearer sound, free of static, than the AM radio dominant at the time. This, however, was the depressed decade of the 1930s and the radio industry was in no mood to take on a new system that required a radical overhaul of both transmitters and receivers. Armstrong found himself boxed in on every side. It took him until 1940 to get a permit for the first FM station, erected at his own expense, on the Hudson River Palisades at Alpine, New Jersey. It would be another two years before the Federal Communications Commission allocated him a few frequencies.

After a hiatus caused by the Second World War, FM broadcasting began to expand. Armstrong again found himself impeded by the FCC, which ordered FM into a new frequency band at limited power. Armstrong's inventions made him a rich man – he held forty-two patents in his lifetime – but he also found himself ensnared in a protracted legal dispute with RCA, which regarded FM radio as a threat to its AM radio business. In addition, several other corporations had also immersed him in legal battles on the basic rights to his inventions, which left him financially drained.

Armstrong's business worries also put a strain on his marriage. One day, during a violent argument, he struck his wife on the arm with a fireplace poker. She went to stay with a relative and Armstrong fell into a deep depression. Beset with personal problems, embittered and drained by years of litigation and facing financial ruin, he chose to commit suicide. On the night of Sunday, 31 January 1954, Armstrong jumped to his death from the thirteenth-floor window of his New York City apartment. An employee found his body, fully clothed with a hat, overcoat and gloves, the following morning on a third-floor balcony. His suicide note to his wife said: 'May God help you and have mercy on my soul.'

Armstrong's widow doggedly renewed her husband's patent fight. In late December 1954 it was announced that an out-of-court settlement had been reached through arbitration with RCA. Following a further series of protracted court proceedings against a coterie of other companies, she was able to formally establish Armstrong as the inventor of FM over five of his basic FM patents. Marion Armstrong ultimately won $10 million in damages. She founded the Armstrong Memorial Research Foundation in 1955, and ran it until her death in 1979 at the age of eighty-one.

For the first few decades after it was invented, the radio was immobile. The wireless used a large amount of energy and was too easily broken because of the fragile valves inside. During the 1940s both those drawbacks were about to be overcome. There was a concerted effort during the Second World War to decrease the size and power consumption of valves, predominantly because the receivers used in radio-controlled bombs depended on valve technology.

In 1947 three research physicists working at Bell Laboratories in America developed the transistor. Walter Brattain, John Bardeen and William Shockley realised that pioneering research into crystals carried out by Russell Ohl a decade earlier could lead to a solid-state alternative to the thermionic valve. The transistor was smaller and more reliable than its cumbersome predecessor. This meant that radios could become smaller and more portable. The initial batch of transistors had used a constituent called germanium as the conducting material. Although it tested well in the confines of the laboratory, it proved too fragile for everyday use. Germanium would eventually be superseded by a silicon alternative a few years later.

The first transistor radio to be produced commercially went on sale in November 1954. It was a joint venture between Texas Instruments in Dallas and Industrial Development Engineering Associates (IDEA) in Indianapolis. Texas Instruments had sought an established radio manufacturer to develop a portable radio but none of the major manufacturers seemed all that interested. However, Ed Tudor, the President of IDEA, jumped at the chance, predicting sales would reach the 20 million mark in three years.

The look and size of the Regency TR-1 received favourable reviews, but comments about its performance were usually adverse. The circuitry had been reduced considerably in an effort to keep costs down and this had severely limited the volume level and sensitivity of the radio. A review in 'Consumer Reports' references the high level of noise and instability on certain frequencies, advising against the purchase.

The radio operated from a 22.5-volt battery and sold for $49.95. Sadly, the high price meant that initial sales were poor and limited to around 150,000 units in the first year. Nevertheless, within a few years prices had fallen sufficiently and transistor radios became very popular. Other companies introduced alternative models and by the end of the decade, almost half of the ten million radios made and sold in America were the portable transistor type.

The birth of radio was a long drawn-out affair. It had a disparate group of inventors and entrepreneurs all claiming parentage but we are no nearer to finding one single person who could claim to be the 'father of radio'.

These men had the marvellous gift of developing their own inventions or adapting other people's theories and ideas. Together they created a superb piece of technology that helped to save lives and influence and entertain others.

By the early 1920s, radio had advanced sufficiently from an invention that was the exclusive preserve of hobbyists and enthusiasts to a medium with real mass-market appeal. Radio was about to hit the mainstream.

Chapter 2

Britain Begins Broadcasting

Alexander Meissner, working for the Telefunken Company, developed the first valve transmitter capable of sending out stable continuous waves in Germany. He prepared a patent application for a 'set for generating electric oscillations' early in 1913. Soon after, the Marconi Company adapted the Meissner Valve Generator to wireless telephony, and obtained a range of 80km for speech transmission.

When the First World War started, all experimentation by wireless amateurs was forbidden. The new broadcast developments were harnessed to the needs of the armed forces and intelligence services. The Admiralty took over the production of equipment at Marconi's Chelmsford works and a wireless interception service was started there.

In 1918, when the war finished, the Marconi Company resumed their development of long-distance radio. They built a two-valve transmitter in County Kerry in Ireland. The Marconi Company chose Ballybunion because of its flat landscape and its direct line to their target: North America. The receiving equipment was installed at Louisburg, Cape Breton, in Nova Scotia.

The main purpose of the tests was to prove that, with the arrangement of the oscillating valve transmitter and the modern Marconi valve receiver, only a small amount of power was needed to relay telephonic or telegraphic messages across the Atlantic. The power of the transmitter was one-sixth of the total power used in the first American transmission tests in 1915. A wavelength of 78.9 kilohertz was used to carry the first European voice across the Atlantic, that of W.T. Ditcham, a Marconi employee.

In Britain wireless instruction courses were given at Crystal Palace and Marconi House. As interest spread, various wireless clubs were formed and members began to build experimental sets. These experimenters, or 'hams', moved away from Morse code transmissions and began to use wireless telephony. At one point nearly twenty stations were transmitting speech and music to the London area on Sunday mornings.

Some of the broadcasters included Harry Walker (2OM) who always introduced his musical programmes with 'This is Brentford Calling!' Bill Corsham (2UV) also gave local talent the opportunity to broadcast, and is credited as the inventor of the QSL card. He would reply to anyone who sent a reception report by sending a card confirming details of the broadcast.

During this period a sizeable number of people were building experimental wireless sets, many using surplus equipment left over from the First World War. Predictably, dishonest dealers would sell ineffective or obsolete equipment that wasn't suitable for the reception of broadcasts to unsuspecting constructors.

People quickly realised the potential for the commercial broadcasting of speech and music. Early in 1920 the Marconi Company began test transmissions from their Chelmsford base. These transmissions emanated from a group of sheds to the rear of Marconi's New Street Works.

The first real telephony broadcast in Britain happened on 15 January 1920 by station GB90MZX. Some of the first transmissions on 109 kilohertz contained readings from Bradshaw's railway timetable. Many appreciative reports were sent to the Marconi Company from European countries. Two of these broadcasts were even heard by the Australian Marconi Company, who reported weak, but solid, signals that were heard in Melbourne, although speech couldn't be distinguished successfully.

The Marconi engineers were greatly encouraged and began refining and modifying their transmission array. From 23 February 1920 listeners heard two daily programmes from MZX, each transmission lasting half an hour and consisting mainly of news, musical items and gramophone records. One of the engineers, G.W. White, was a talented pianist and he recruited an oboe player, fiddler and clarinetist from within the Marconi works. Eddie Cooper, another employee, sang with local bands and was invited to perform a few songs. His powerful tenor voice transmitted well so he was asked to recruit some more local musicians.

One such musician was Miss Winifred Sayer, a local soprano, who became the first British female to broadcast. Miss Sayer was a member of the clerical staff of a neighbouring factory, Hoffmann's, and so would have to be paid. A payment of 10s (ten shillings) per performance was approved and a series of fifteen-minute transmissions were arranged. She sang in the main transmission room amidst all the noisy equipment. She recalled singing unaccompanied on three separate occasions, including a duet with Eddie Cooper, holding an adapted telephone receiver to her lips.

Winifred only realised the full impact of what she had done when she finished her debut broadcast. As she left, Godfrey Isaacs, Marconi's managing director, told her she had made history. A few weeks later she received a copy of the Marconi Company's *Souvenir of a Historical Achievement* booklet.

Many radio amateurs and ships' radio operators tuned in to the transmissions, and reports were received from as far away as Belgium, Norway and Portugal. This greatly encouraged Marconi and he was persuaded to broadcast the world's first live recital by a professional musician, the legendary Australian diva, Dame Nellie Melba.

Lord Northcliffe, the proprietor of the *Daily Mail*, arranged the event. However, the prima donna seemed reticent and it took a personal intervention by Northcliffe to eventually persuade Melba to participate in the broadcast. She maintained that her voice was 'not a matter for experimentation by young wireless amateurs and their magic play boxes.'

She was finally persuaded to appear once her list of conditions was met. One unusual demand that Dame Nellie insisted upon was that before she broadcast she must have her favourite meal, which consisted of chicken with unleavened white bread washed down with champagne.

Before the broadcast Dame Nellie was shown around Marconi's factory. As part of her tour she was shown the 140-m-tall twin towers that would transmit her recital. Her guide explained to her that from the top of the mast her voice would be heard throughout the world. 'Young Man,' she retorted, 'if you think I'm going to climb up there you are very much mistaken.'

Rather wisely they didn't place her in the noisy transmission room. Instead Melba sang in a small ex-packing shed well away from the equipment. In the run-up to the transmission, Marconi staff did everything they could to improve the appearance of the shed where the broadcast was to take place. The walls were whitewashed but the surroundings still remained fairly basic and littered with equipment. A thick carpet had been laid over the stone floor to help with the acoustics, though Dame Nellie didn't seem too keen on it and asked for it to be removed.

The transmitter was warmed up just before the broadcast was due to begin, but just as Dame Nellie was about to sing, a photographer in the transmitter room decided to take a last-minute photograph. As he did so, the engineer on the control panel saw the reflected flash and panicked. He hurriedly switched everything off, which meant that the long, involved process of transmitter run-up had to be started from the beginning again.

At 7.15pm she finally began to perform, using a microphone created with a telephone mouthpiece and wood from a cigar-box. The whole contraption was suspended from a modified hat-rack by a length of elastic. She started her recital by singing 'Home Sweet Home'. Melba was only supposed to sing three songs but was persuaded to sing four more and closed with the National Anthem.

Within days the Marconi Company had received enthusiastic letters from all over the world. They received excellent reception reports from Spain, the Netherlands, Norway, Sweden, Germany and Sultanabad in Northern Persia. It has been suggested that the signal was received so strongly at the Eiffel Tower in Paris that gramophone records were made. Dame Melba described the event as being the 'most wonderful experience of my career' and took a great personal pride in being the first singer to broadcast all over the world. She was not the only opera legend to sing at Chelmsford, however: the famous contralto Dame Clara Butt also made an appearance at the Marconi Works later on.

The engineers found it particularly difficult to balance the sound of musical items. Altering the settings to suit one instrument would often directly affect the tone of another. Musical instruments were especially challenging to capture in a natural manner as fiddles sounded like oboes and pianos sounded shrill and piercing. This would be especially difficult if the instrumentalists were accompanying a singer. Reducing the treble would often make baritones and deep tenors sound muffled, and duets were virtually impossible due to the technical limitations of the microphone.

The first complete wireless receivers appeared in the UK in 1920. They were made by BTH: The British Thomson Houston Company. They were joined a year later by Burnham & Company and Leslie McMichael Limited. Receivers in those days were extremely unselective, and speech and music required a fairly large space on the longwave band.

Broadcasts continued until the autumn, when there was a temporary setback. The Postmaster-General informed the House of Commons that the Chelmsford experiments had caused several cases of interference with other stations, notably those broadcasting military communications. One such incident reported that a pilot was crossing the English Channel in a thick fog. While trying to obtain weather and landing information from Lympne in Kent, all he could hear was music.

The Chelmsford broadcasts had also impeded the Post Office's newly opened arc transmitting station near Oxford, which carried press and Foreign Office Morse transmissions to America and India. Given reports such as these the Post Office had no alternative but to withdraw the Marconi Company licence.

Although MZX Chelmsford was forced off the air, the number of amateur transmitting and receiving licences continued to increase. By March 1921 the Post Office revealed it had issued over 150 transmitting licences and over 4,000 receiving licences with another 1,700 requests waiting to be processed.

In the USA commercial broadcasting was a great success. There was a large demand for receivers, and by 1922 numerous stations were broadcasting. This fuelled interest in Britain and manufacturers fervently wanted to emulate the success of their American counterparts.

In August 1921 the Marconi Company was permitted to broadcast calibration signals in Morse each week for a period of half an hour. This at least informed the listener which frequency he was tuned to but it could hardly be described as captivating listening.

Then, in December, a petition signed by sixty-three wireless societies with over 3,000 members was presented to the Post Office demanding the resumption of regular telephony broadcasts. Early in 1922 the Postmaster General permitted the Marconi Company to resume a programme of speech and music lasting fifteen minutes to be included in the weekly half-hour transmission of calibrations.

So the Marconi Company began the first ever officially approved broadcast service to Britain. 2MT, based at Writtle, would be the first UK station entirely devoted to radio telephony. The station was set up in an old shed on the edge of a five-acre field that was liable to flooding, though the engineers quite liked this as the flood plains actually helped with their ground wave aerial experiments.

The 'studio' was essentially one end of the shed cleared of everything except a microphone and piano. The station's transmitter was powered by an army petrol-driven engine and fed the signal to the aerial slung between two high masts. This would become a standard arrangement for the early stations.

Shortly before the first broadcast, a glass condenser completely shattered and a loud crackling sound indicated that the signal had been lost. The faulty component was hastily replaced and the transmission was able to proceed.

That debut broadcast consisted of three five-minute sections of Morse code followed by brief intermissions when the station closed down to listen for emergency messages. At about 7.35pm one of the engineers stepped to the microphone and introduced a selection of recordings by Robert Howe. The records were played on a mechanical gramophone with the microphone held near to the horn speaker. Writtle had perfected what Eckersley called the 'ultimate volume control'. This basically consisted of the engineer moving the microphone closer or further away from the gramophone horn.

The broadcast went smoothly enough, but the engineers all agreed that the inaugural broadcast hadn't been up to scratch. The signal was badly muffled and any speech had an abnormal tone. Preliminary reception reports were reasonable, but far from enthusiastic.

After two or three more transmissions in the same vein, Writtle began to receive numerous complaints about the poor signal quality. They eventually discovered that the glass condenser that had been hurriedly replaced on the opening night was to blame. The faulty component was replaced and within a fortnight the complaints had dwindled away.

As mentioned previously, the wonderfully unconventional Peter Eckersley helmed these experimental broadcasts. He'd had considerable experience in radio communications by this time. Eckersley had served as a wireless equipment officer during the First World War with the Royal Flying Corps. He was then sent to the Wireless Experimental Station at Biggin Hill where he conducted experimental work on duplex telephony for aircraft.

After the war Eckersley joined Marconi's Wireless Telegraph Co as the head of the experimental section of the aircraft division. During his time in this role he designed the Croydon Airport ground station transmitter. On the merger of the aircraft division and the field station section, he became head of the design department.

Peter Eckersley was quite happy to stay away from the microphone for the first few broadcasts. He preferred to go home and listen to the transmissions from his home a few kilometres away. However, one night in March 1922, he decided to stay behind and get involved with that evening's broadcast.

Fortified by a good meal and a couple of drinks at the local pub, Eckersley was in a relaxed, playful mood and decided to have a go in front of the microphone. He adopted a less formal tone than his predecessors, which made for a livelier broadcast than the usual bill of fare served up to that point. He conducted the whole show more or less single-handedly. He overran, failed to play all the records and neglected to shut down for the regulation three-minute station breaks. Then, at just after nine, he even began to sing. His presentation style was widely welcomed by the listeners and he became the regular announcer from then on.

These early transmissions at Writtle consisted mainly of gramophone records, but a live concert featuring Miss Nora Scott was broadcast on 11 April 1922. The concert brought appreciative reports from the north of Scotland to the south of England; it seemed that the whole country wanted to listen.

The celebrated Danish tenor Lauritz Melchior also made his debut on 2MT a few months later. His inexperience of the new medium was to cause problems for the Writtle engineers. Accustomed to projecting his voice to fill a large opera house, Melchior took a deep breath and began to sing: loudly! The resultant cacophony overloaded the transmitter, blew the circuit breakers and put the station off air. In the end they positioned him 4.5m from the microphone before obtaining an acceptable level. For months afterwards, the engineers always used to refer to any faulty or damaged component as being caused by the 'Melchior breakdown'.

Lauritz Melchior was the pre-eminent Wagnerian tenor of his era but he chose to sidestep any lengthy numbers by the German composer. Instead he sang several short pieces, including the Danish and British National Anthems. At one point strong atmospheric interference was encountered, but it didn't mar the enjoyment of the event too much. The concert was immediately followed by several speeches from dignitaries such as Queen Alexandra of Denmark and Guglielmo Marconi.

Fans in Melchor's native country were able to hear the broadcast even if they didn't have a wireless set. A signal from the receiving station in Denmark was fed to the telephone exchange in Copenhagen.

This made it possible for anyone to listen to the 'overseas' broadcast, all for the price of a phone call.

Eckersley's team at Writtle was a small, enthusiastic bunch of individuals. They included Basil McLarty who, along with Harry Kirke, designed the transmitter. Then there was Edward Trump, Freddy Bubb and the honourable Rolls Wynn. Elizabeth Beeson handled the secretarial duties. She was the daughter of the landlord of 'The Cock and Bell', which was the local pub frequented by the engineers.

2MT's broadcasts became more ambitious and the station even produced a piece of drama. The balcony scene from *Cyrano* was chosen as it's staged in semi-darkness with virtually stationary actors and so Eckersley considered it highly suitable for broadcasting. For economic reasons, the engineers would make up the majority of the cast.

The leading lady boasted the marvellous sobriquet of Uggy Travers. Uggy, whose real name was Agnes, had been called in to help the engineers with their lines. For this broadcast the cast would need to be sat around a table, speaking their words into a solitary microphone passed from hand to hand as the lines demanded. To avoid any disasters on the night, the engineers rehearsed this procedure beforehand in Rolls Wynn's house. With the script resting on the table, each person read their lines into the back of a large spoon, which was doubling as the microphone. They had to practise passing the 'microphone' from actor to actor without dropping it, or avoid making a noise by rustling their scripts.

The Writtle transmissions had led to the creation of a sister station in London. The Marconi Company had been given another licence to operate a transmitter at Marconi House, in London. This was the famous '2LO', which started broadcasting on 11 May 1922.

The Marconi employee in charge of the London station's output was Arthur Burrows. His background was journalism, but he had helped train radio operators during the First World War. Burrows was widely regarded as a friendly and affable man, but he was determined to avoid the chaotic manner in which 2MT was run. Under his leadership the London station was run more professionally than its counterpart at Writtle.

Burrows had a keen eye for detail and preferred everything within his purview to be clearly defined. His staff had clear guidelines to adhere to. However, it could be said that his edicts resulted in more mundane programming emanating from 2LO than the equivalent ones from Essex.

Burrows initiated an elaborate signing-off procedure for each broadcast. A set of tubular bells suspended from the ceiling would be chimed sequentially when the station closed down. As a cost-saving measure, it was decided to rent the bells instead of purchasing them outright. Burrows had considered the purchase price of £20 as excessive.

The station planned an ambitious insert to the inaugural broadcast by covering the Ted 'Kid' Lewis-Georges Carpentier boxing match at Olympia in West London. The commentary was relayed live from the ringside via telephone to Marconi House. Unfortunately, the relay had to be curtailed abruptly when Carpentier won by a knockout within the first minute.

On 21 June 1922, 2LO broadcast a moving ceremony to commemorate the 348 employees of the Marconi companies who were killed during the First World War. The managing director, Godfrey Isaacs, unveiled a memorial to wireless operators who were mostly employed by the Marconi International Marine Company and were lost at sea.

The initial broadcasts were speech only but permission was eventually granted to broadcast music. The first musical item was transmitted on 24 June 1922 and presented by Beatrice Eveline on the cello, Ethel Walker on piano and a baritone, Charles Knowles. The microphone was placed in the centre of the studio and the performers were assembled strategically around it at varying distances so that no one voice or instrument would overpower the others.

The musical content of the early 2LO programmes ranged from chamber music ensembles to solo artists and singers. At one point they even had the massed Band of the Irish Guards assembled in the small studio for a performance, though it's not known if they were able to fit the entire band into the cramped surroundings.

Stanton Jeffries was appointed Musical Director and there followed a series of regular concerts culminating in four performances daily during the first 'All-British Wireless Exhibition and Convention'. This event took place in the Horticultural Hall in Westminster between 30 September and 7 October 1922 and provided many with their first encounter with radio. 2LO's output was relayed via a powerful valve set with large speakers placed around the hall.

Also in September the station provided daily reports of progress in the 'King's Cup Air Race', and on 20 October broadcast its first comedy programme, *A Cockney Fragment from Life*, written and performed by

Helena Millais as 'Our Lizzie'. Also in October the Prince of Wales's speech at the British Boy Scout Rally at Alexandra Palace was relayed to the listeners.

That station's power was 1.5 kilowatts on a frequency of 840 kilohertz. Programmes were initially broadcast for just an hour each day and the London station could only transmit for seven minutes at a time. This was because the 'operator' had to listen on the wavelength for three minutes for possible instructions to close down. This restriction was finally lifted on 8 January 1923.

Marconi wasn't the only company interested in the new medium. In Manchester the Metropolitan Vickers Company Limited commenced test broadcasts on 16 May 1922 from its own station, identified as 2ZY. The broadcasts originated from the firm's base in Old Trafford.

Metropolitan Vickers concentrated mainly on heavy engineering, but it did have a sister company called The Radio Communication Company. The RCC specialised in marine communication and they supplied most of the start-up equipment. The man placed in charge of operating the station was H.G. Bell, who would later become chief engineer and manager of the Stretford and District Electricity Board.

The tests were organised by Kenneth Wright, an engineer in the Research Department of Metropolitan Vickers. Tests were sporadic at first and 2ZY didn't have a regular studio until July 1922. The engineers experimented with new and more sensitive types of microphones. At first, carbon granule microphones were used, the same kind used by 2LO. Later experiments used 'photophone' microphones, which employed the principle of reflecting light onto a selenium cell. Photophone microphones were better for picking up the sound from musical instruments, but they were considerably larger and more cumbersome than the carbon granule microphones so were quickly removed from service.

The transmitter was installed on 25 October and tests began shortly after. The regular tests relied heavily on gramophone records and a deal was struck with the Gramophone Company to provide these free of charge. This was on the understanding that they would receive free publicity and no other company's records were used. The first live concert was broadcast on Halloween 1922 and used musical talent from within Metropolitan Vickers staff.

Birmingham would also get its own station. The Midlands was the perfect opportunity for the Western Electric Company to enter the fray. The company had played a large part in expanding the British telephone network and had also developed a popular public address system. They had installed an experimental station at their third-floor laboratory at Oswaldestre House in Norfolk Street, London, which had been allotted the call sign 2WP. The various technologies used by these systems meant that they were in a good position to enter the broadcasting field.

Nevertheless, they had procrastinated more than Marconi and Metropolitan Vickers so they had to move fast if they were to be ready by the proposed start date of November 1922. Frank Gill, the president of the Institution of Electrical Engineers, was employed to supervise the transfer of 2WP's transmitter and equipment from London to Birmingham and a fleet of steam-lorries transported everything to the new site. Western Electric also secured an agreement with the General Electric Company to house its station within their works at Witton.

Once everything had been delivered to Birmingham, the engineers managed to equip the studios and commence tests within three days. The station utilised a new type of microphone used in Western Electric's public address system. The 'double button' carbon microphone increased the frequency response and was far superior to other types of microphones at the time. It contained a stretched steel diaphragm that could receive the highest notes of a soprano without causing any distortion.

2LO's engineers were anxious to check that the new Birmingham station would not interfere with their broadcasts so the two broadcast simultaneously on 5 November. No interference was subsequently reported.

Although the stations in London, Birmingham and Manchester were initially independent of each other they were unlikely to stay that way for long. The Post Office had come under severe pressure to allow national broadcasting and so they met with representatives from interested groups on 18 May 1922. At this point there had been twenty-three requests from applicants to start broadcasting. The Post Office asked them to come up with a mutual scheme for broadcasting but five months of discussions resulted in no proper proposal. At each meeting there was a large amount of conflict as each company vehemently protected its own interests.

The delay in reaching any agreement was criticised by the government and the press. Finally, however, on 18 October 1922, a decision was made to form a single entity that would be responsible for broadcasting in Britain.

The British Broadcasting Committee was a transitory body, which lasted until the new company could be officially registered. This transitional period lasted until December and early broadcasts were announced as 'on behalf of the Broadcasting Committee'.

The efforts of the committee resulted in the formation of the British Broadcasting Company, which would establish a nationwide network of radio transmitters, many of which had originally been owned by member companies from which the BBC was to provide a national broadcasting service.

The British Broadcasting Company Ltd was incorporated under the 1908 to 1917 Companies Acts with a share capital of £100,000 made up of 99,993 cumulative ordinary shares valued at £1 each. The six largest companies provided the capital: these were Marconi, BTH, GEC, Western Electric, Metropolitan Vickers and the Radio Communication Company. The director of a seventh firm, Burndept Ltd, was co-opted onto the board to act on behalf of the smaller shareholders.

This British Broadcasting Company would derive some income from a licence sold to listeners, with the rest coming from the manufacture and sale of licensed receiving sets. Initially there weren't enough licences available so in many cases a personal letter was sent out giving permission to listen. These letters later enclosed a permit to be carried at all times.

In the early days there were different types of licences. The broadcast licence fee was introduced on 1 November 1922 and cost 10s. This applied to listeners who purchased one of the official BBC sets. The constructor's licence was introduced in October 1923 at a cost of 15s. This was for amateurs wishing to construct their own receivers using British-made components only. This licence was only available briefly and was terminated in July 1924. The 'Receiving Licence' was introduced in January 1925 and all previous forms of licences were abolished. These licences covered all wireless sets in your home, though you were required to have a separate licence if you had a car radio.

Half the licence-fee went to the BBC and the rest to the six manufacturers mentioned previously. Receivers manufactured in the UK were required to carry a distinctive label indicating that royalties had been paid.

In addition the listeners had to pay two tariffs. The first was based upon the various components in the receiver and went to the BBC. The second was a levy of 12s 6d per valve-holder, which would go to the Marconi Company as a royalty in return for allowing their patents to be used.

On Tuesday, 14 November 1922, the BBC officially took control of 2LO and started broadcasting in the medium waveband. It did so without an official licence from the Post Office. The licence was eventually issued retrospectively in January 1923. The broadcasts emanated from the seventh floor of Marconi House on the Strand. Initially the power was 100 watts on 857 kilohertz.

The first broadcast consisted solely of a news summary and a weather report read by Arthur Burrows, although he never announced his name on air. This was a deliberate BBC policy that was to be adhered to for many years. The bulletin was read at 6pm and repeated again at 9pm. He read each bulletin twice, once quickly and once slowly. Burrows then asked listeners to say what they preferred.

The first bulletin included details of a speech given by the Conservative leader, Bonar Law, the opening of the Old Bailey sessions, the aftermath of a 'rowdy meeting' involving Winston Churchill, a train robbery, the sale of a Shakespearean first folio, fog in London, and the latest billiards scores. Due to pressure from the newspaper industry, the BBC was not allowed to transmit its news bulletins until after 6pm. This was so that newspaper sales would not be lost to a BBC radio news service.

Early broadcasts included news, household hints and talks. The BBC's first talk took place on 23 December 1922. The subject matter was never recorded, but we do know the topic of the second: 'Listeners in' to the London station on 27 January 1923 were treated to a thrilling talk entitled 'How to Catch a Tiger'. Arts reviews and literary discussions began in February and were aired weekly.

The first chairman of the British Broadcasting Company was Lord Gainford, a former Postmaster General. The new organisation didn't have a general manager at its helm during its inception but eventually appointed John Charles Walsham Reith as its first general manager on 14 December 1922.

Reith, a minister's son, had been born in Stonehaven in the North-East of Scotland in 1889. His family moved to Glasgow while he was an infant. After his education at Glasgow Academy he spent two years studying at the Royal Technical College in the city. He served an engineering apprenticeship at the North British Locomotive Company where he specialised in radio communication. In 1913 he found a job with a firm at the Royal Albert Dock in London.

At the outbreak of the First World War Reith joined the Fifth Scottish Rifles and gradually rose up the ranks. He transferred to the Royal Engineers as a lieutenant. In October 1915, while fighting in Flanders, he was badly wounded by a sniper's bullet, which pierced his cheek and left him with a noticeable scar. He spent the next two years in the United States, supervising armament contracts for the Ministry of Munitions. He was promoted to captain in 1917, before transferring to the Royal Marine Engineers as a major the following year. He returned to the Royal Engineers as a captain in 1919.

Back in Civvy Street, he went to Glasgow to work as general manager of an engineering firm. He then returned to London in 1922 and started working as secretary to the London Conservative group.

His strict Presbyterian upbringing had a big influence on his life and greatly affected his demeanour. The dour, humourless Scot described his early life when interviewed for John Freeman's *Face-to-Face* programme in the 1950s. Reith portrayed a life of austerity and in his own words he 'never learned that life was for living'.

Reith was a difficult man to like as he tended to be quick-tempered, and steadfast in his orthodox attitudes. Nevertheless, he turned out to be the perfect man for the job. He had honed his business skills during his spells in America and Glasgow. He knew nothing about broadcasting but relished the challenge his new position presented him. His managerial ability would prove particularly useful as he challenged the petty rules and regulations that the bureaucracy imposed on his new company.

Reith immediately began innovating, experimenting and organising. Under his leadership the BBC was turned from what was a handful of experimental stations run by wireless manufacturers into a cohesive organisation with a mission to 'inform, educate and entertain'. He directed staff to make quality broadcasts with a strong moral underpinning.

The other stations in Birmingham and Manchester also transferred operations to the BBC. The Birmingham station should have been allocated the call signs 2BM or 2BH, but it was thought these may have been confused with Bournemouth therefore arbitrary letters were chosen for the call sign. It was a similar situation with the Manchester station:

2ZY should have been allocated 2MC, or even 2MT, but the Marconi Company already held these so an alternative was chosen at random

The Birmingham station, now renamed 5IT, was due to commence broadcasting at 17.20pm on 15 November. Invitations to local dignitaries were duly sent but this presented the staff with a big problem. The studio and offices were extremely cramped and it would have been difficult to fit everyone into the confined space. Fortunately, the Birmingham weather offered a timely solution. A thick fog had enveloped the city on the opening night and a number of those who were invited to the ceremony decided to stay away.

The opening broadcast was very successful and received many favourable press comments. The transmission had been received in Paris and Copenhagen as well as across the Midlands. Yet behind the scenes things had not gone so well. The power generator had begun to overheat and one of the engineers had to spend the entire evening in the basement with a grease gun

Frederick Percy Edgar was the general manager and opening announcer for 5IT. Edgar was a former music hall act who became a director of a concert agency in Birmingham. He was approached initially by the Western Electric Company to supply artists to perform on the radio station, and was surprised to receive an offer shortly afterwards to manage the station itself. His friends and colleagues advised him to turn the job down as radio would be just a passing craze, but he accepted and became the BBC's senior regional director.

One hour and forty minutes after Birmingham commenced broadcasting, 2ZY went on the air in Manchester. This was polling day in the 1922 general election and all three BBC stations stayed on air until 1am to carry results phoned through from Reuters. There had been prolonged dialogue regarding news and it was agreed that Reuters would supply a summary for use after 6pm. This was dictated over the telephone to Marconi House and passed again by telephone to the other stations.

5IT pioneered many innovations, from employing the first full-time announcers to launching children's programmes. *Children's Hour* was the brainchild of the station's chief engineer, A.E. Thompson, who broadcast as 'Uncle Tom'. The practice of children's announcers adopting the appellation of 'Aunt' or 'Uncle' was also adopted at other stations. Published stories were forbidden due to copyright so Thompson invented stories about a cat called Susan, based on a grotesque china cat he found in a junk shop. Another engineer, F.H. Amis, who became 'The Fairy Dustman', joined him later.

Many music hall acts were given the opportunity to broadcast. Thompson recalled one act from the Aston Hippodrome who didn't fully grasp the concept of sound broadcasting. The woman flounced around the studio as if on stage with Thompson trailing behind her holding the microphone on a long lead. After this incident a small podium was built out of crates and the performers were told to stand on it and not move.

General Electric's premises at Witton were unsuitable for the BBC's long-term aims and alternative accommodation was sought. In August 1923 the Birmingham station moved to the second floor of a large building at 105 New Street. The new premises included a studio, control room and offices. There were also two other rooms set aside for a reception and band room. The transmitter was moved to a power station at Summer Lane.

During the summer of 1923, 2ZY in Manchester also moved to a more central location, on the fifth floor of a warehouse in Dickinson Street, though the old Metropolitan Vickers' staff thought that the move from Trafford Park meant the loss of the 'family spirit' that pervaded the old premises.

The more centrally placed premises, however, were useful in enticing artists who didn't want to go to Trafford Park. The new studio was adjacent to the Manchester Corporation Power Station where the aerial mast was erected and the close proximity of studio, transmitter and aerial mast negated the need for any GPO landline. The move to Dickinson Street coincided with Kenneth Wright's departure for a senior post in London and a replacement station manager was sought.

2ZY's new station manager, Dan Godfrey Junior, created an orchestra of twelve players known as the 2ZY Orchestra. He also initiated an opera company and there began a regular series of live music broadcasts. Many works, principally by British composers, were given their first broadcast performances by the 2ZY Orchestra: these included Holst's *The Planets*, Elgar's *Enigma Variations* and *The Dream of Gerontius*. The 2ZY Orchestra was eventually renamed the Northern Wireless Orchestra in 1926.

The Newcastle station 5NO started broadcasting in December 1922. The BBC moved into brand new premises in the city centre.

The annual rent for 24 Eldon Square was £250 per annum. This was the first time that a BBC station had not benefited from existing studios used by previous test stations.

The transmitter site was located a mile away from 5NO's studios at the stable yards of the Co-operative Wholesale Society in West Blandford Street. The BBC had rented one of the stables to house the transmitter and a nearby factory chimney supported the aerial.

The first day of broadcasting was set to be 23 December 1922. At the last moment, however, technical problems arose when trying to connect the transmitter to the studio, and the cast and crew of the opening broadcast were forced to broadcast from the stable yard instead. The engineers wheeled empty carts into the yard and placed chairs on them and microphones were connected to the nearby transmitter. Tom Payne, the station director, made the opening announcement and played his violin. There followed several songs from May Osborne and a cello recital from James Griffiths.

The studio link was ready the following day and this is officially considered to be the opening date of 5NO. There was a slight problem during the first broadcast from the studios when a howling dog, kennelled nearby, could be heard faintly in the background.

Tom Payne's tenure as station director was brief and Bertram Fryer replaced him in March 1923. Fryer adopted the best features of other stations. As 'Uncle Jack' he initiated the Fairy Flower League and encouraged youngsters to develop an interest in animals and plants. Fryer also installed receivers with loudspeakers in local cinemas so patrons could listen to 5NO while watching a silent film.

The arrival of these new stations signalled the end for 2MT, British radio's pioneering radio station. While its transmissions had lasted for just under a year, its impact was immense. Often a one-man show, but always a team effort, 2MT established an individuality all its own. But they were never incorporated into the BBC and 2MT suspended transmissions on 17 January 1923. At the final closedown, the engineers toasted their listeners' good health with water doubling as champagne. They used a pop gun to simulate the popping of the cork.

They were never ordered to cease broadcasting and the 2MT licence was never formally revoked, but they had been eclipsed by various other, more powerful, stations and the Writtle team had realised it was time to stop and move on to new projects. Basil McLarty moved from Writtle to head the Design and Installation department, while Harry Kirke became head of the BBC Research Department. Rolls Wynn eventually became the BBC's Chief Engineer.

The first BBC outside broadcast took place on 8 January 1923. Post Office engineers installed a quarter-mile-long lead-sheathed cable between the Covent Garden Opera House and Marconi House. Mozart's *Magic Flute* was transmitted to an extremely appreciative audience. Announcer Stanton Jeffries was placed in the prompter's box to convey what was happening during silent periods on stage. The operas *Hansel and Gretel, Pagliacci* and *Siegfried* were all transmitted on subsequent evenings. The season climaxed on 17 January with a performance of *La Bohème* starring Dame Nellie Melba. Harrods remained open that night so customers could listen to the event in the Georgian Restaurant.

The first experimental transatlantic radio relay took place in the early hours of 26 November 1923. The BBC used eight stations throughout Britain for the tests, all relaying the same programme from London. For the first fifteen minutes nothing was heard, then faint, unintelligible speech was heard. A telegram was sent to the BBC asking for a piano solo. Three minutes later the notes of a piano were heard, followed by the words, 'Hello America'. A month later, on 30 December 1923, the first continental broadcast to the UK was made by landline from Radiola in Paris.

The BBC's station in London continued to innovate. The first religious address on the BBC was broadcast on Christmas Eve 1922. The Reverend John Mayo, the Rector of St Mary's Whitechapel, broadcast two sermons that day. The first one was broadcast during a programme for children and he returned later in the evening to speak to the adults.

In April 1924 the BBC decided to broadcast a regular religious service and chose St Martin-in-the-Fields Church as the venue. This soon became a regular monthly feature, with Dick Sheppard, 'the radio parson', sharing his pulpit with other clergy.

Certain sectors of the entertainment industry felt threatened by radio. The Theatrical Managers Association withdrew its support as it feared the new medium would affect ticket sales. Other impresarios agreed and prevented their artists from participating in broadcasts. The embargo stayed in place throughout the BBC's early years.

The Newspaper Publishers' Association also viewed the BBC with suspicion. During the spring of 1923 John Reith received an ultimatum from the association warning him that if the corporation didn't pay a hefty fee, none of the association's publications would carry radio listings. Although the embargo was short-lived, it gave Reith the idea of publishing a dedicated listings magazine. And so the first edition of the *Radio Times, the official organ of the BBC*, duly appeared on 28 September 1923.

Radio Times was a joint venture between the BBC and publisher George Newnes Ltd, who produced, printed and distributed the magazine. The *Radio Times* quickly established a good reputation. It used the leading writers and illustrators of the day and the covers from the special editions of this period are now regarded as design classics. Weekly sales had reached 750,000 by the end of 1924 and the BBC assumed full editorial control in 1925.

BBC English

The clipped tones that the announcers generally spoke in became known as 'BBC English'. This had previously been known as Received Pronunciation or Oxford English. In fact the BBC claims they never imposed a standard accent on its broadcasters and newsreaders. Their only stipulation was that people around the country should be able to understand the announcements. It's more likely that 'BBC English' was a consequence of the limited social strata from which BBC employees were drawn, rather than a matter of deliberate strategy.

Although the BBC didn't promote a standard accent they did decide on the pronunciation of individual words and places. Executives compiled lists of proper pronunciations for use by the announcers. They were particularly keen to pronounce place names correctly so they wrote to many local dignitaries across the country to ascertain the correct enunciation. Unfortunately, the local worthies often disagreed about how to pronounce the names of the locality they lived in.

2LO's initial broadcasts were estimated to have reached an audience of about 18,000. The BBC's home was still Marconi's headquarters in the Strand. Eventually it was decided to seek suitable accommodation of its own. It settled on the Institution of Electrical Engineers' building in Savoy Hill, near the Embankment. The premises were opened on 1 May 1923. The first studio was built on the third floor and by 1926 there were five studios in use.

The atmosphere at Savoy Hill was friendly, but idiosyncratic. Each evening a rat-catcher looked for vermin and office boys in rubber gloves puffed germicide everywhere to prevent coughs and sneezes among the broadcasters.

The engineers were keen to avoid the acoustic problems that dogged transmissions from Marconi House. Each studio floor had a thick wall-to-wall carpet and the walls were lined with five layers of canvas. Each layer was stretched on a wooden frame and placed an inch apart. A sixth layer of yellow net curtains completed the acoustic barrier.

Taking centre stage in the Savoy Hill studios was a brand new microphone. The 'Magnetophone' or 'Marconi Round Sykes', weighed 25lb and was trundled around on a wheeled base. The microphone was enclosed in a wire mesh and was likened by the engineers to 'a meat safe on a large tea trolley'. The contraption was almost 1.5m tall and required a row of car batteries to power it.

Henry Joseph Round, Marconi's Chief Engineer, designed the new microphone. Round had joined the Marconi Company in 1902, shortly after Marconi made his transatlantic wireless transmission. He was sent to the USA where he experimented with a variety of different aspects of radio technology.

The First World War broke out in 1914 and Round was seconded to Military Intelligence with the rank of captain. He established a chain of direction-finding stations along the Western Front, which proved so successful that another set was installed in England. For his services during the war, Round was awarded the Military Cross. When the Second World War broke out in 1939, the British government again called on his services. This time he was involved in developing Sonar.

The inaugural broadcast from Savoy Hill featured speeches by Lord Birkenhead, Sir William Bull and Lord Gainford. Lord Birkenhead caused some concern by failing to appear at the appointed time. Enquiries revealed he was dining next door at the Savoy Hotel and a deputation was hastily despatched to get him. When he finally appeared he seemed 'unsteady on his feet' but delivered his speech perfectly.

The Band of the Grenadier Guards provided the music and entertainer Norman Long was on hand to provide light relief. Norman Long was the first entertainer to be 'made' by radio, making his debut on the BBC on 28 November 1922. Long's music hall slogan 'A Song, a Smile and a Piano' was changed to 'A Song, a Joke and a Piano', on the basis that you can't hear a smile.

By May 1924 every available space in the Institution of Electrical Engineers' building was utilised so they expanded into the adjacent Savoy Hill Mansions, part of which had been destroyed in a Zeppelin raid in 1917.

The BBC continued to expand. The first new station to start broadcasting in 1923 was 5WA Cardiff, which commenced transmissions on 13 February. It proved difficult to find suitable premises and compromises had to be made. The only location that could be found was in cramped rooms above a cinema across from Cardiff Castle.

The studio location made soundproofing difficult and the noise of trams trundling by on the street below was a permanent backdrop to broadcasts. The transmitters were located a mile away at an electricity sub-station in Ninian Park Road. Rex Palmer was the station director but his tenure at Cardiff was brief as he was appointed to a similar role at 2LO in April.

A month later, on 6 March 1923, Glasgow got its regional station when 5SC started transmissions (the initials in the call sign 5SC being derived from 'Scotland'). Even though it was located in Glasgow, 5SC's 1.5-kilowatt transmitter was designed to ensure it could be heard far beyond the city limits. The new BBC station was 150 times more powerful than that of the temporary Daimler station that preceded it.

The great and the good were in attendance at its inaugural broadcast. The sound of bagpipes was followed by BBC general manager John Reith's opening announcement. Reith then introduced BBC Chairman Lord Gainford, the Principal of Glasgow University and Glasgow's Lord Provost, who all made short speeches. These were interspersed by several musical pieces performed by members of the Royal Carl Rosa Opera Company and the studio orchestra.

Early employees at 5SC included Mungo Dewar, Alexander H.A. Paterson and Kathleen Garscadden. These three performed administrative duties in addition to on-air duties. They were also regulars on *Children's Corner*, which began on 5SC's second day. For this programme, Kathleen Garscadden chose to be known as 'Auntie Cyclone'. However, London didn't share the joke and soon forbade her to use this nickname.

They believed using meteorological terms could confuse the children, and so she simply became 'Aunty Kathleen'.

Glasgow's headquarters consisted of several cramped rooms in an attic at Rex House, 202 Bath Street. It may have lacked space but 5SC enjoyed the latest technology. The studio boasted a Western Electric double button carbon microphone and amplifier.

As the BBC had no engineering staff of its own, they drafted in engineers from the Marconi Company to erect the transmitter. The Marconi Q type transmitter was situated at the Port Dundas Power Station in Edington Street. The location was regarded as exceptionally suitable for the BBC's needs. It was located on one of the highest points in Glasgow and the nearby canal would provide a good 'earth'. The aerial was slung between the second-floor wireless room and one of the two chimneys, which towered above the complex.

The citizens of the North East of Scotland joined the radio age on 10 October 1923. Herbert Carruthers, 5SC's station director, made the arrangements for Aberdeen's opening night broadcast. The Aberdeen radio station was assigned the call sign '2BD' and began broadcasting on 606 kilohertz.

The launch of 2BD was to be an important one for the BBC as it was planned to be a simultaneous broadcast between Aberdeen and London. If things went to plan, the company could draw together programmes originating from different locations and thereby eliminate the need for stations to be autonomous. By May 1923 simultaneous broadcasting was possible between main transmitters and relay stations, but the quality was not felt to be high enough to provide a national service or regular simultaneous broadcasts.

On the opening day things didn't look promising. A severe gale had swept the country and the trunk telephone wires supplying the north of the country were down. The engineers worked all day to ensure that the array of post office lines required to carry the programme southwards would supply a trouble-free connection.

At 6.50pm the sound of pipe music was heard, followed by station announcements read by station director R.E. Jeffery. He had been the principal actor in many important touring companies before running the Aldwych Theatre in London. There followed a relay from London before reverting back to the Aberdeen studio. The opening night programme, including the official opening ceremony, was simultaneously broadcast to all BBC stations between 9 and 9.30pm.

The BBC rented accommodation at the rear of Aberdeen Electrical Engineering's property at 17 Belmont Street. Access to the premises was gained by the narrow stairway at the rear of the shop. On the second floor were a couple of small offices and a large room, which was an old meeting hall that overlooked the main Aberdeen–Inverness railway line. The area was converted into a rudimentary studio by draping heavy black curtains on the wall to deaden the noise from the passing trains. The studio was particularly affected by the vagaries of the North East Scotland weather. No central heating or air conditioning meant the temperature veered between freezing cold in the winter to boiling hot during the summer.

2BD's transmitter was located in the premises of the Aberdeen Steam Laundry Company in Claremont Street. Perhaps this was taking the idea of Steam Radio too literally. From there the signal was sent to the aerial, which was strung up between two tall Marconi masts. It was perhaps unwise to site the studios near to several electrical generators as these often interfered with the signal. Despite the low power, the initial broadcasts were heard in Norway, and 2BD's output was clearly picked up in the United States during International Radio Week in November 1924.

Bournemouth (6BM) opened seven days after the Aberdeen station on 17 October 1923. The purpose-built studio premises were located at Holdenhurst Road and the transmitter was situated nearly 3km way at North Cemetery, Bushey Road. 6BM became the first ever BBC station to broadcast a 'commercial'. Following a concert by the Bournemouth Symphony Orchestra, its founder, Sir Dan Godfrey – whose son was the station director at 2ZY in Manchester – encouraged listeners to 'come to Bournemouth and hear the Orchestra perform live'. Godfrey was heavily reprimanded for 'advertising' on the air.

Each station was responsible for its own output, provided they adhered to the guidelines issued from London. To help fill the broadcast hours, the London office would often send up a large leather trunk filled with sheet music and material for the local actors and musicians to perform. A team of peripatetic performers also travelled around the regions doing one-night stands at each location. Their gruelling fortnightly schedule would take them to stations throughout Scotland, Northern Ireland, Wales and England.

To give a flavour of the type of output these early stations provided, let's look at some of the programmes that 2BD in Aberdeen transmitted.

Technical limitations meant certain types of programmes were more prevalent. The single person talk was a necessary requirement of the early days in broadcasting because there was only one microphone in the Aberdeen studio. The microphone's gauze structure gave the technical boffins loads of headaches, especially in summer. Bluebottles frequently got inside from the bottom of the stand and became trapped in the mesh. When this happened, all the listener would hear was a loud humming noise and the broadcast would have to be suspended while the offending insect was removed.

Station 2BD became one of the first stations to broadcast a weekly fifteen-minute sports programme. Peter Craigmile, an international football referee, would preview the week's forthcoming events. The Aberdeen station was also responsible for another broadcasting first when it transmitted a Gaelic language programme in 1923.

The 2BD Repertory Company was established to perform adaptations of the classics as well as numerous offerings in the local vernacular. These were mostly one-act plays with a handful of characters. One popular drama series was centred on a fictitious castle in Aberdeenshire. 'The House at Rosieburn' told the tale of a witch burnt at the stake who, before she died, placed a curse on the inhabitants. One actor, playing a witchfinder in the series, was the envy of his colleagues when he was given an official sanction to utter the line, 'Come on, you old bitch!' Although quite tame by today's standards, this was considered quite outrageous language at the time.

The station had its own twelve-piece orchestra, formed in 1924, to provide musical entertainment. In accordance with BBC policy, the musicians would dress in full evening wear for every performance. The only time they were permitted to remove their jackets was on particularly hot days during the summer. There was no air conditioning in those days and they couldn't open a window because of the noise of the nearby railway. The performers had no choice but to sweat it out.

Their double bass player once unwittingly put the station off the air during a heatwave. Perspiring profusely, he looked down and noticed a plug in a nearby wall socket. In a fit of pique he bent down and unplugged what he thought was the radiator. Unfortunately, it turned out to be the socket for the one and only microphone and it was several minutes before the error was realised.

In addition, popular music hall artistes were invited to perform when they were in the area. Mabel Constanduros was a character actress who was incredibly popular in the early days of wireless. She performed comedy monologues as various members of the Buggins family. Mabel appeared before the 2BD microphone on 3 February 1928 in a programme with the catchy title *Mrs Buggins gives a party in the Aberdeen Studio.*

The establishment of 6BM in Bournemouth fulfilled the BBC's initial obligation to build eight stations in the main areas of population. It needed to cover the widest population in the quickest way possible in order to make the service profitable for the company's shareholders. The obvious solution was to keep adding to the existing stations, but, unfortunately, they couldn't increase their power without causing mutual interference.

The Covent Garden Opera broadcasts had demonstrated it was possible to use a post office landline over short distances. Therefore the BBC wondered if it would be possible to relay a programme from one part of the country to another. They decided to experiment and hire two lines from Birmingham to London. They used one for communication between the studios and the other to convey the broadcast itself, and on 20 March 1923 a musical programme was relayed from Birmingham to London. The experiment was repeated on 16 April, only this time from Glasgow. The engineers were pleased with the quality even though there was some distortion on high notes.

The signal needed to be boosted so that the simultaneous broadcast would reach outlying transmitters. In order to do this a series of amplifiers would be used in each location. Two further tests were carried out using this method, but these were less satisfactory as severe interference was heard. Eventually the problems were ironed out and the era of simultaneous broadcasting had arrived.

The new technique also enabled the news to be broadcast directly from London. This ended the frustrating nightly practice of phoning all the provincial stations and dictating the news script to them. John Reith read the first simultaneous news broadcast on 29 August 1923.

Simultaneous broadcasting was seen as a way to broadcast special programmes to a wider audience. It caught on quickly and nearly a third of 2LO's output was being relayed to the provincial stations by the end of 1923. This figure had risen to fifty per cent by the time 1924 had concluded.

When new stations were introduced they were classed as either a 'main' or 'relay' station. The relay stations would receive programmes from the nearest main city studio via telephone circuits, but each station had a studio to opt out of the main transmissions if required. The first relay station was Sheffield (2FL) which first transmitted on 16 November 1923. The Sheffield station was located in the Union Grinding Wheel Company premises in Corporation Street. Several other stations joined the list in 1924: 5PY in Plymouth first broadcast on 26 March from Athenaeum Lane and the 100-watt transmitter was situated at the sugar refinery in Mill Street.

In January 1924, four months before the launch of the Edinburgh station, it was estimated that there were around 800 radio enthusiasts in the city. 2EH started relaying BBC programmes on May Day 1924. The Edinburgh office was located in the back premises of Townsend and Thomson's music shop at 79 George Street. The studios were very cramped so the opening ceremony was staged at the Usher Hall. The Lord Provost of Edinburgh, Sir William Sleigh, formally declared the station open. During the ceremony, a local news bulletin was read from the front of the stage, including the story of a major fire that had broken out in Edinburgh that evening. The local fire chief happened to be sitting in the audience and, on hearing the news, hurriedly left the hall to attend the blaze.

As a relay station, it mainly rebroadcast transmissions from London and other stations, but also made a number of its own local programmes. Initially 2EH broadcast every afternoon with a complete evening programme once a week. This was originally on Wednesdays but then transferred to Fridays some time later.

At first the afternoon programmes comprised mainly of music, which was relayed from St Andrew Picture House one afternoon a week, with another afternoon of music provided via a wireless link from Daventry. Then there was the 'Pianoforte Trio', a group of well-known local musicians. They played three times weekly at first, but after a few months they were allowed to broadcast on a daily basis.

A short half-hour religious service was broadcast from the studio every Sunday. From 1925, this was done regularly on the second Sunday of each month from St Cuthbert's Parish Church. There were also outside broadcasts covering the proceedings of the General Assemblies of the Church of Scotland and the United Free Church of Scotland.

2EH relayed several outside broadcasts during this period. They covered the proceedings when the prime minister at the time,

Stanley Baldwin, was given the Freedom of the City in May 1926. Probably the most complex of these events happened on 14 July 1927. This was when the Prince of Wales attended the opening of the Scottish War Memorial. Ten microphone points were wired up and a team of engineers was sent from head office to oversee the broadcast.

Like all the other BBC relay stations, 2EH operated on a low power of just 200 watts. This was around one-tenth the power of a main station and posed a special problem for the engineers when Edinburgh's unique geography was factored in. Finding an ideal transmitter location that would serve most of the city proved difficult. Eventually the transmitter was installed in a wooden building in the quadrangle at Edinburgh University. The 47-m aerial was suspended from a chimney located at the back of the medical buildings as this was thought to be the most centrally situated position available.

After complaints from listeners about poor reception, the BBC sent a development engineer to Edinburgh University to fix the problems. Once it was confirmed that the aerial was not at fault, the report concluded that the presence of shielding metallic rocks was the cause of the poor signal strength. Also a proper earth couldn't be attained at the site because of the local rock foundation. These problems, among others, led to 2EH moving to larger premises at 87 George Street on 31 July 1925. The transmitter was also eventually moved because the Edinburgh University site was unavailable. Just after midnight on 26 April 1931 the transmitter was dismantled and re-erected 1.5km away in the bakery at St Cuthbert's Co-operative Society headquarters in Port Hamilton. The job was completed by 3pm the same day, just in time for the afternoon programmes.

Relays of 2ZY were introduced during the summer. These covered Liverpool (6LV), Leeds and Bradford (2LS), Nottingham (5NH), Hull (6KH) and Stoke-on-Trent (6ST). 6LV began broadcasting on 11 June 1924. The studio was above a café in Liverpool's Lord Street while the transmitter and engineers were located on the first floor of a disused paint shop near Smithdown Road. Looking through 6LV's programme listings, it seems that Gaillard and his orchestra were regularly relayed from the Scala Cinema in Lime Street during those early days. 6LV was notable for being the first provincial station to broadcast a speech by the ruling monarch.

2LS opened on 8 July 1924 with an outside broadcast from the stage at Leeds Town Hall. Station director Philip Fox introduced the Lord Mayors of Leeds and Bradford, the first time that the civic leaders had appeared on the same platform. The first regular programme, *Children's Hour*, appeared the following day. 2LS arranged concerts from the large towns across Yorkshire: Huddersfield Nights and Harrogate Nights being just a couple of examples. It was also the first station to broadcast a concert from the bottom of a coal mine at Normanton on 28 November 1924. The programme featured a comedian and music from the Whitwood Colliery Silver Prize Band.

6KH started on 15 August 1924 from studios in Bishop Lane in Hull's Old Town. The transmitter and aerial was located on a high building in the industrial area of Wimcomlee. Local shows included talks and lectures, sports commentaries and light music from the bands that played in local cafés, such as the Hammond's Café Trio and the Powolny's Restaurant Bijou Orchestra.

6ST first transmitted from small premises in Stoke-on-Trent on 21 October 1924. The studio was laid out in the standard BBC way at the time with large yellow drapes around the wall to keep any extraneous noise out. There was an adjacent control room that was only big enough to hold three people. The transmitter was housed in a wooden hut next to the Trent and Mersey Canal with the aerial slung from a nearby chimney stack. John Snagge, the well-known radio announcer and television commentator, started his BBC career at 6ST as an assistant station director.

2DE started to relay the Aberdeen programmes to Dundee on 9 November 1924. It operated from the former offices of the jute manufacturers Gilroy Brothers in Lochee Road. In the days when 2DE had its own exclusive wavelength there were reception reports from as far away as the south coast of England. 2DE's transmitter was attached to the highest chimney of the highest jute factory in Dundee: Caldrum Works in St Salvador Street. The aerial was a large cage of cables over 70m long, which formed the radiator. The length of cable had hoops along its length to spread the wire out and give a constant diameter to the cage. This made the structure look a bit like a long sausage, so it was often known as a sausage aerial. This attracted considerable public attention when the big sausage was slung from the chimney to a pole.

Finally, 5SX relayed 5WA's programmes to Swansea, commencing on 12 December 1924, from the top floor of the Oxford Buildings on Union Street. Alderman John Lewis, Mayor of Swansea, was in attendance as

Wilfrid Goatman made the opening announcement. Goatman had joined the BBC just four days before and recalled that the premises still needed a lot of work done to them. There was no soundproofing in the main studio and the sound of traffic passing by frequently interfered with the broadcasts.

These relay stations were an unhappy compromise. Financial constraints meant that this was the easiest, and most economic, way for the BBC to cover the main areas of population. However, intense rivalries between cities and towns in a large geographical area meant that these small stations pleased nobody. Listeners in Dundee felt they had nothing in common with their counterparts in Aberdeen and resented the programmes of 2BD being foisted upon them. There was a similar situation with listeners in Sheffield, Cardiff and Edinburgh who all begrudged predominantly hearing programmes from their feeder stations. A few of them decided a more suitable compromise would be to broadcast programmes from London instead.

Because of wavelength limitations, the BBC opened only one new main station in 1924. The governor of Northern Ireland officially opened 2BE in Belfast on 15 September from its premises in Linenhall Street. The station broadcast on a wavelength of 689 kilohertz and the first voice to be heard on the airwaves was that of Tyrone Guthrie. The station's programming was limited and had a relatively small audience.

The British Broadcasting Company Ltd didn't sell airtime for commercials. However, its licence did allow for it to carry sponsored programming and eight such sponsored broadcasts were aired in 1925. Yet its main income was still derived from the sale of radio sets and the licence fee. By the autumn of 1924 the GPO had issued over 1 million receiving licences. The BBC had twenty radio transmitting stations in operation and 465 employees.

The original stations reached about half of the population with signals that were strong enough to be received by a crystal set. The 'listeners in', as they were called, had to show a remarkable dedication to listen to these early broadcasts. In the 1920s buying a radio was a very expensive proposition and only the rich would have been able to buy a commercially built radio.

The alternative to buying a commercially made set was to build a home-made radio. The simplest receiver was a crystal set, which used a mineral crystal as a rectifier, and enabled the listener to tune in the station on headphones. This was assuming they were within range of a station and could locate a receptive part of the crystal using a wire probe called a 'cat's whisker'. Attaching a second or third pair of headphones would decrease the signal strength even more. A novel way to allow several people to listen would be to place the headphones in a pudding basin. This would amplify the volume sufficiently for other people to hear provided they were close enough.

The crystal set was the most popular way of listening during the pioneering days of wireless broadcasting. Its popularity was not surprising since it was cheap to buy and simple to operate. The main disadvantage of owning a crystal set was that it wasn't very good at separating two or more stations transmitting on wavelengths that were close together. It was this inadequacy that eventually led to the crystal set's demise.

Although the weak signals generated by a crystal set were usually only strong enough to satisfy one listener wearing headphones, it was possible to buy special crystal set amplifiers, to boost this volume to a point where it was powerful enough to drive a loudspeaker. These amplifiers were usually powered by a six-volt battery, which would last for many months due to the very low current consumption. A circuit using a differential microphone, a reed and a magnet helped to attain this increased volume. Provided the listener was living within a twelve-mile radius of one of the BBC's main stations, it would be good enough to make family listening comfortable.

A crystal set had no power from mains or batteries and relied on the energy from the radio waves that were collected in its large long-wire aerial to work. The aerial for a crystal set would need to be many tens of metres long for anything to be heard and would have to be fixed to the roof or erected out in the garden. A long piece of wire dangling from a chimney stack and tied to a washing pole with empty cotton reels providing insulation at either end made an ideal aerial. If the listener had no garden or space to erect his aerial, an indoor frame aerial could be used as an alternative. The listener's licence allowed the use of up to 30m of aerial wire.

To obtain greater range or volume, a valve receiver was needed. This would amplify the radio signals and the sounds so that weaker signals would be heard. A valve radio would require more electronic components and both low voltage and high voltage batteries for it to work, making it a more expensive proposition. A valve cost a week's wages (for the average worker) and was extremely fragile and easily damaged, and needed frequent replacement. They were liable to burn out after only a short time because of the great heat needed to drive them. The valves were also known as 'bright emitters' as their filaments lit up like electric lamps.

A large high-tension battery and a low-tension accumulator provided the power. They could be recharged like a modern mobile phone or laptop battery. The difference was, these beasts weighed several pounds and were full of corrosive acid so a slow careful walk to the local dealer was a necessity. Most radio owners had two, one to use and one being charged. You could easily spot a wireless enthusiast in those days by the acid burns on his hands. Wireless dealers, as well as cycle shops and garages, would recharge batteries and accumulators for about 6d.

One person could listen on headphones, but a loudspeaker was needed for a family to 'listen in'. These were usually metal or wooden horns fixed to a telephone receiver. These horn speakers were eventually replaced by the moving coil type, which appeared just before 1930. This is the type we still use today.

The units containing the receivers were often skilfully made of French-polished walnut or mahogany with intricate details adorning the cabinet. Most orders for wireless cabinets were sent out to cabinet-makers employed in the furniture industry, which explains the high level of craftsmanship. Needless to say this quality was only evident on the most expensive sets. For budget sets, oak and even cardboard were used as the cabinet's material.

The early valve receivers were mostly of the tuned radio frequency type, which often had a confusing array of different switches and knobs on the front panel. Tuning in a station required the operation of numerous controls, and to change waveband sometimes required the insertion of different tuning coils. Many receivers had a reaction control that adjusted the sensitivity and selectivity of the receiver. The receiver would oscillate and act like a transmitter if the reaction control was adjusted too far, consequently interfering with everyone else's receiver in the neighbourhood.

The modern type of receiver appeared in 1925. The super-heterodyne, or superhet for short, was a highly sensitive and selective radio, with a vastly reduced selection of tuning controls. In those days mains electricity was provided by hundreds of small companies, which supplied electricity in a variety of different voltages, frequencies and even DC (Direct Current). The national grid was established in the early 1930s and standardised the supply. This made it much easier to produce mains power supplies for receivers. It also helped to establish mains-powered radios. Mains-powered receivers and battery eliminators first appeared in 1926. The first receiver to entirely dispense with batteries was the 'Baby Grand', made by Gambrell Brothers Limited.

For many years, the terms 'wireless' and 'radio' were used to describe the same thing. The difference being that 'radio' was the American version of the British 'wireless'. The receiver was called a 'wireless' because there were no wires linking it to the transmitting station. It was also called a 'radio' because it was a receiver of radiated electrical signals, both terms being equally precise. Increasingly, however, the American term was embraced in Britain as it came to mean something more modern than the seemingly slightly old-fashioned term 'wireless'.

During this period many people couldn't afford to buy a radio for the home, but fortunately there was an affordable alternative available to potential listeners in major towns and cities. For a weekly fee, usually a shilling, subscribers to wired relay networks received the BBC services without the need for a wireless set.

The first service of this kind appeared in Hull in 1928 and proliferated over the next three decades. The companies who provided these systems usually had a mast with the necessary equipment to receive the radio signals off the air. These masts were often at an elevated site that afforded good reception. The signals were then diffused via a cable network to homes in the area. At its height, radio relay services were used in over one million homes throughout the country.

Subscribers simply had a loudspeaker that could be switched on and off and a dial that could select one of the three BBC radio programme services. The loudspeaker required no battery or mains power, and in many households it would often be left on all day long.

The BBC broadcast the chimes of Big Ben for the first time in December 1923. A couple of months later another time-related tradition was introduced. The Greenwich Time Signal was first broadcast in February 1924. The 'pips', as they're more generally known, were the brainchild of Frank Dyson, ninth Astronomer Royal. The six-pip time signal was devised in discussion with Frank Hope-Jones, inventor of the free pendulum clock, who had initially advocated a five-pip signal. The sixth pip signals the start of the next minute.

Two mechanical clocks located in the Royal Greenwich Observatory, which had electrical contacts attached to their pendula, originally controlled the pips. Two clocks were used in case of a breakdown. The clocks sent a signal each second to the BBC, which was then converted to the audible oscillatory tone used in the broadcast. Before the BBC started using the pips, a pianist in a studio would play the tune of the Westminster chimes, synchronising the 'bong' with the clock in the studio.

On 15 January 1924 the BBC broadcast its first specially commissioned 'listening play', as it was dubbed. *A Comedy of Danger* by Richard Hughes was about a group of people trapped in a Welsh coal mine. To help with atmosphere the listener was encouraged to turn out the lights and listen in the dark. The producers encountered several problems when it came to sound effects. The acoustics of the studio couldn't recreate a vast cavernous mineshaft so the actors had to place their heads in buckets.

A male voice choir was enlisted as the script called for 'distant snatches of hymn-singing'. However, the men were very enthusiastic, and once started nothing could stop them. The producer put them in the corridor outside with a soundproof door he could open and shut when required.

An explosion proved to be trickier as any large noise would have disturbed the transmitting equipment and put the BBC off the air. Therefore, instead of a loud bang, the listeners heard a muffled thud. This was not what a group of invited journalists heard. The gentlemen of the press were ushered into a private room to listen to the play and so at the appropriate point an engineer concocted a substitute explosion in the room next door. The 'explosion' gained enthusiastic approval from the press. They never discovered they had heard it through the wall.

Another 'first' happened on 23 April 1924 when the first broadcast by a reigning monarch was transmitted. Millions heard King George V open the Wembley Empire exhibition. Traffic was stopped on Oxford Street as crowds gathered to listen on loud speakers.

The Theatrical Managers Association's embargo was eventually lifted and many music hall acts auditioned for the BBC. Not every act was suited to the new medium, however, and many found it hard to perform in an empty room with no interaction from the audience. Many famous music hall acts failed to make the transition.

Those acts that could adapt found that it boosted their careers enormously. Willie Rouse, who came to be known as 'Wireless Willie', was extremely adept at improvising. His spontaneous gags were usually at the expense of the announcer or accompanist. Rouse was immensely popular but his radio career was short-lived. Wireless Willie died in 1928 at the age of fifty-one.

John Henry had failed to make it big in the theatre but he found his niche in broadcasting. Henry specialised in monologues delivered in a droll Yorkshire accent. His idiosyncratic tales usually featured his wife, 'Blossom', and his dog, 'Erbert'.

Some of the early stars of radio were rather unusual in that they didn't come from a show business background. Sir Walford Davies introduced the first schools broadcast in 1924 and embarked on a series of regular talks on music. His talks were knowledgeable but never opinionated and Davies had a sense of intimacy that remains the key to successful broadcasting to this day.

Sir Oliver Lodge had already occupied a key position in the history of radio. Lodge was well known in radio history as the inventor of the 'tuned circuit' but his ability to explain difficult scientific subjects clearly and simply endeared him to the audience. He was totally relaxed in front of a microphone and would often pause while searching for the correct word or phrase. After being introduced, Lodge would clear his throat and proceed with his talk. This throaty growl became his 'signature tune'.

A.J. Allen was another regular contributor of talks who seemed to be making it up as he went along. In reality, his pauses were cleverly contrived to give the impression of spontaneity. In reality, all his talks were well rehearsed and nothing was left to chance. It later emerged that A.J. Allen was the pseudonym of Leslie Lambert who held a top-secret post in the government.

In 1925 the London masts were moved to a new more powerful transmitter at Selfridge's Department Store in Oxford Street. The transmitter was housed in a large hut on the roof and the aerial was supported on two 38-m pylons. The transmitter and masts remained at this site until 1929.

Early in 1926 a talk by Father Ronald Knox caused some controversy. *Broadcasting from the Barricades* featured fictitious reports of rioting in

London by the unemployed. Many listeners panicked and thought that Britain was in the throes of revolution. The longest running programme in the history of British radio made its first appearance at this time. *The Week's Good Cause* was first broadcast on 24 January.

The BBC was still in its infancy when the values laid down by John Reith were first put to the test. The organisation clashed with the government over editorial independence during the general strike. In May 1926 Britain's miners went on strike and, in a move of solidarity, other industry workers joined them. For nine days the nation's industry was at a standstill. At such a politically sensitive time the company had to tread carefully.

Radio was the only widely available source of news, as many of the newspapers were not published. There were barely any means of communication between authorities and the general public. The Conservative government published its own 'British Gazette', which was launched and edited by the then chancellor, Winston Churchill. He could see that radio was a more direct and adaptable medium and he lobbied Prime Minister Stanley Baldwin to commandeer the company.

John Reith argued that such a move would destroy the company's reputation for independence and impartiality. He put forward a persuasive defence and Baldwin ruled that the BBC should remain independent. This judgement displeased Churchill who complained bitterly about the decision. He later said, 'It was monstrous not to use such an instrument to the best possible advantage.'

Reith argued later that the Conservative, Labour and trade union perspectives had all been reported impartially. Despite that, the TUC and the Labour Party, led by Ramsay MacDonald, disagreed. They said the BBC refused airtime to their representatives. The BBC's reporting of the strike was guarded and far from comprehensive but is now regarded by historians as reasonably fair.

Reith didn't recoil from dealing with politicians. He labelled Neville Chamberlain as 'cavalier' because he supposedly didn't take the power of broadcasting seriously enough. During the 1929 election he had invited the main party leaders to make election addresses on the BBC. Reith thought Ramsay MacDonald was ineffective but Stanley Baldwin impressed him by asking about the social make-up of the radio audience.

Reith's own election performance turned out to be a substandard affair. The BBC's election results service had been carefully rehearsed, but during the second bulletin of the evening Reith announced he would read all the results himself. He hadn't been doing it long when listeners rang asking for the reader to be clearer and to read slower. When advised of this, Reith curtly replied, 'I will not read any slower; I am going on announcing.'

The general strike had been an early test of the BBC's resolve to remain editorially independent and it happened when the organisation was in a period of transition. On 5 March 1926 a Parliamentary Committee led by Lord Crawford published its broadcasting report. The committee recommended that broadcasting should be conducted by a public corporation 'acting as trustee for the national interest'. It called for the termination of the British Broadcasting Company and the creation of a Crown chartered, non-commercial organisation.

The Crawford Committee also approved the general tone of John Reith's programming approach. Not surprisingly, he became the first director general of the BBC when the company assets were dissolved and it became the British Broadcasting Corporation on 1 January 1927. At this point the original staff of four had grown to over 500. The newly formed corporation boasted ten main transmitting stations, ten relay stations and over two million listeners.

Chapter 3

The British Broadcasting Corporation

On 20 December 1926 the Crown charter and licence agreements creating the new Crown chartered, non-commercial organisation were published. On 31 December 1926 the contracts of 773 British Broadcasting Company Ltd staff were terminated and, with the dissolution of the company, shareholders were paid at par value. The company assets were transferred to the British Broadcasting Corporation. At this point the wireless manufacturing companies ceased to be directly responsible for broadcasting in this country.

While the BBC was no longer an independent commercial company, the aim of the charter was that it would stay free of central government intervention and an appointed Board of Governors would oversee the corporation. John Reith, the director general, was honoured with a knighthood in December 1926. Under his leadership the BBC continued its mission to 'inform, educate and entertain'.

The Republic of Ireland also began state broadcasting around this time. The Department of Posts and Telegraphs licensed 2RN in Dublin, which began broadcasting on 1 January 1926. The station's name was said to stand for 'To Eireann,' which was derived from the last three syllables of the song *Come Back to Erin.* Initially the station broadcast three hours a night Monday to Saturday with a two-hour broadcast on Sundays.

The studio for 2RN was originally located at Little Denmark Street in Dublin before moving, in 1928, to the GPO building on Sackville Street, which had previously been The Wireless School of Telegraphy. The first Irish broadcast, a Morse code message declaring Irish independence, had been transmitted from there during the Easter uprising in 1916.

Reception of the station for many listeners outside Dublin was problematic so the Cork broadcasting station 6CK was established in 1927. The studios of 6CK were in a section of the old Cork City Prison, which had only been recently vacated, having been used as an overflow prison for political prisoners at the end of the Irish War of Independence. This station relayed many of 2RN's programmes, as well as contributing programmes of its own to the network.

The Dublin and Cork stations were boosted by a high-power station at Athlone in 1932. At that time the Dublin, Cork and Athlone stations adopted a common call sign of 'Radio Athlone'. 'Radio Eireann' replaced this call sign five years later and lasted until 1966 when its name was changed to RTE Radio (Radio Telefis Eireann) to reflect the new governing authority established in 1960.

Back in the United Kingdom, by 1927 the General Post Office had issued 2.3 million receiving licences and the BBC had twenty-one stations broadcasting around the nation. Nearly all of the country could obtain at least one of these, even with low-quality receivers. Some eighty-five per cent of the population could receive a choice of programmes.

The year 1927 saw a number of 'firsts': the first live sport broadcast was transmitted on 15 January when Teddy Wakelam commentated on the rugby union international between England and Wales; a week later the first football match was broadcast.

Christopher Stone presented a record programme on 7 July 1927 and became the first British disc-jockey. His relaxed, conversational style was at odds with the BBC's usual starchy form of presentation. Stone's programmes became highly popular and lasted until 1934, but he was barred from the BBC as a consequence of signing up for a show on Radio Luxembourg. Stone's fee for this show was £5,000, a princely sum at the time.

The BBC upheld its mandate 'to inform, educate and entertain' when it assumed responsibility of the Promenade Concerts in 1927. This British tradition had been held at Queen's Hall in London since 1895. The aim of the 'Proms' was to promote classical music to a wider audience: it did this by adopting a less formal approach and offering reduced ticket prices. Despite this, the enterprise was encountering financial difficulties until the BBC stepped in.

Some believed that broadcasting the concerts would reduce audience numbers, but this fear proved groundless, and the BBC continues to broadcast the Proms to this day. For the first three years the concerts were given by 'Sir Henry Wood and his Symphony Orchestra', until the BBC Symphony Orchestra was formed in 1930.

Peter Eckersley, the driving force behind the Marconi 2MT transmissions, eventually became the first chief engineer of the BBC. He was well liked by colleagues and management and appeared to be doing a first-rate job. However, that changed in 1929 when he became embroiled in a divorce scandal.

The married man began an affair with Dorothy 'Dolly' Clark, the estranged wife of the BBC's music adviser, Edward Clark. It seems that Edward Clark maintained a relaxed attitude regarding the affair. Clark and Eckersley even visited Germany on BBC business and took Dorothy with them. Later, Eckersley got bolder, taking Dolly on a BBC trip to Brussels and was repeatedly seen with her at Savoy Hill. Colleagues who knew his wife, Stella, were incensed and in January 1929 the matter was brought to the director general's attention.

It's widely believed to this day that John Reith went ballistic and fired Eckersley on the spot due to his deeply held religious beliefs. The truth is that Reith's response was a little more measured. Reith invited Eckersley and his lover to his flat to explain that the chief engineer would lose his job unless the affair ended. Reith's wife, Muriel, broke the news to Stella Eckersley, who knew nothing of the affair.

Reith informed the BBC Board of Governors. Some of them demanded instant dismissal but the chairman and Reith were more guarded. They wanted Eckersley to resign and be reinstated later if his marriage could be saved. Reith even asked for advice from the Archbishop of Canterbury and Commissioner of the Metropolitan Police, Admiral Byng. The archbishop's views were not revealed but Byng, who was far more insistent, said a BBC man was like a policeman and never off duty. He recommended that Eckersley should be dismissed.

Reith didn't follow the commissioner's advice. Eckersley promised to end the affair and offered his resignation, but Reith recommended to the governors that he should stay. The matter looked as if it had been resolved, but it proved to be only a short-term respite.

In April 1929 Eckersley met Dolly again at a party and immediately decided to divorce his wife. He proffered his resignation to the Board of Governors. When the governors accepted the chief engineer's resignation, they awarded him £1,000 in appreciation of his valuable service and retained him as a consultant for another year.

After the BBC he got involved with stations in Europe and was also involved in a company called Broadcast Relay Service Limited, which provided an early form of cable radio. It was Eckersley who suggested the company change its name to Rediffusion. He began to wire parts of England to receive continental transmissions, but this was curtailed at the intervention of the GPO.

From 1937 onwards Eckersley worked for British Military Intelligence MI6 to help combat propaganda coming from abroad. This was despite his association with the British Fascist movement and his wife's links to the Nazis. The couple regularly took holidays in Germany and attended the Nuremberg rallies of 1937 and 1938. The marriage was not a happy one and the couple had separated just before the start of the Second World War. Dolly Eckersley worked for the German broadcasting service during the war. She was tried in 1945 for providing support to the enemy and sent to prison for a year. Peter Eckersley died aged seventy-one on 18 March 1963.

Eckersley was replaced by his assistant of three years, who had also followed him from the pioneering days at Writtle. Noel Ashbridge was a modest man with a solid technical background. His common-sense approach earned him the respect of colleagues and a knighthood in 1935. Shortly after his award, he was appointed as the first technical director of the BBC, a position he would hold until his retirement in 1952.

Another BBC magazine *The Listener* made its first appearance on 16 January 1929. Its published aim was to be 'a medium for intelligent reception of broadcast programmes by way of amplification and explanation of those features which cannot now be dealt with in the editorial columns of the *Radio Times*'. It previewed major literary and musical broadcasts and reproduced written versions of broadcast talks. It also reviewed new books, and printed a selected list of the more intellectual broadcasts for the coming week.

Despite their popularity, the days of the local stations were numbered. It was no longer feasible for twenty BBC stations to continue on twenty different wavelengths. The initial network was proving wasteful and resources were being squandered by duplicating the same kind of programmes on each of the different stations. In 1924 it was felt that technical standards had improved enough for London to start to provide the majority of the output, cutting the local stations back to providing items of local interest.

Simultaneous broadcasting was possible between main transmitters and relay stations, but it was felt that the quality wasn't good enough to provide a national service or regular simultaneous broadcasts. Then, in June 1924, a longwave transmitter (call sign 5XX) was opened in Chelmsford that was intended to spread the wings of the BBC onto foreign shores. BBC management quickly realised that a sustained service could be supplied to most of Britain.

The experimental longwave station was deemed a great success and so a permanent site was sought. Eventually the longwave transmitter was moved to a more centralised location on Borough Hill in Daventry, Northamptonshire. The new station retained the call sign 5XX and was launched on 25 July 1925. The powerful 25-kilowatt transmitter on 187 kilohertz enabled improved coverage across the UK, allowing the BBC to provide a service programmed from London for the majority of the population.

The director general, John Reith, opened the new station by reading a poem called *Danetree* by Alfred Noyes. This referred to a local legend that the oak tree, which stood on top of Borough Hill, marked the exact centre of England. The tree was felled to make way for the main transmission complex but the BBC planted a replacement in a location close to the building.

5XX conducted its first experimental stereo broadcast of a concert in Manchester. The 5XX longwave transmitter beamed the right-hand channel and all the local BBC medium wave transmitters broadcast the left-hand channel.

The BBC opened a central control room in Piccadilly Gardens, Manchester in 1929. Many network radio programmes were produced or broadcast from there. In addition, plays and concerts were staged in an old converted repertory theatre hall in Hulme, which was renamed *The Radio Playhouse*.

The problem of interference from abroad was also becoming serious. In America the random proliferation of stations had become unworkable. In March 1925 an international conference was convened to prevent 'a chaos of the ether'. The conference was to regulate the frequencies and power used by each European country and the Geneva plan was eventually implemented in November 1926.

The Geneva agreement extended the medium wave frequencies up to 1200 kilohertz and implemented 10 kilohertz channel spacing.

Unfortunately, the agreement halved the number of medium wavelengths used by the BBC at that time. Following this, some of their relays in different parts of the country had to share a common frequency.

Two further international frequency revisions were made in 1929, extending the medium wave band to 1500 kilohertz and abandoning the 10 kilohertz channel spacing. In this plan frequencies were assigned to countries instead of individual transmitting stations.

The development of Post Office landlines between studios meant that a sustaining service from London could be provided to the outlying stations. The need and the impetus to make one hundred per cent local programmes faded.

Simultaneous broadcasting made it possible for the development of both the National Programme and the Regional Programmes. On 21 August 1927 the BBC opened a high-power medium wave transmitter at Daventry. The BBC had built a small self-contained building on a site further up Borough Hill to house the new transmitter at a cost of £10,000. It was estimated that the running costs would be £3,000 per annum.

5GB replaced the existing local stations in the Midlands. The new transmitter required a power of between 30 and 50 kilowatts, bigger than any previously built. This allowed 5XX, the experimental longwave transmitter, to provide a service programmed from London that would be available to the majority of the population.

5GB was deemed to be a great success and there followed the establishment of seven regional services across the UK, each broadcasting programmes from its own local studio. 'The Regional Scheme' combined the resources of the local stations into one station in each area, with a sustaining service from London, and the BBC hoped to increase programme quality while also centralising the management of the radio service.

The regions covered were the Midlands, West, North, South East, Scotland, Wales and Northern Ireland. Some local studios were retained to provide programming from particular areas within each region, but these were eventually replaced by the BBC Regional Programme, which was introduced from 9 March 1930.

Also on that date the BBC National Programme was introduced. This supplanted the earlier experimental station 5XX. Its broadcasting hours were from 10.15am until midnight Monday to Saturday, with Sunday's transmission commencing at 3pm.

The new service didn't have a dedicated news department until 1934. Even then it was only used to edit and broadcast news material from other news agencies. News was not permitted to air until at least 6pm each day. This was to fulfil an agreement between the BBC and the newspaper publishers to ensure people would buy a daily newspaper.

With the development of both the National Programme and the Regional Programmes, the smaller local stations slowly died out, and with them went the experimental phase of broadcasting. These changes were also accompanied by changes in terminology. Transmitters were no longer referred to as stations and studios were to be called offices.

This new Regional Scheme required the BBC to build new and more powerful transmitting stations that could carry both the National Programme and the Regional Programme services to the whole country. The first of these 'Twin Wave' stations to be built specifically for the Regional Scheme was Brookman's Park in Hertfordshire. This site was capable of providing signals to London and the South East and so the transmitter at Selfridges was decommissioned. The station was a huge enterprise, using four large lattice towers; two towers were used to support the aerial system for each service.

The Brookman's Park station opened on 21 October 1929 using wavelengths of 1149 kilohertz for 5XX at 70 kilowatts and 842 kilohertz for the Regional Programme at 40 kilowatts. Because the National Programme used shorter wavelengths (higher frequencies) the range was somewhat less than that of the Regional Programme. However, the longwave transmitter at Daventry also transmitted the National Programme and would fill in any areas of poorer reception. Some transmitters also carried the BBC National Programme on a local frequency to supplement the longwave broadcasts from Daventry. Scotland, for example, received an amended service known as the 'Scottish National Programme'. The BBC ensured that the new transmission arrangements would provide robust reception for listeners with both valved radios and humble crystal sets that were still being used.

On 15 January 1934 another international frequency plan was implemented, resulting in more changes to BBC frequencies. This meant that many transmitters had to swap frequencies and the longwave transmitter had to move to 200 kilohertz.

On 6 September 1934 the BBC opened a new Twin Wave station near the town of Droitwich, south of Birmingham, to serve the Midlands region. Droitwich was to replace the existing transmitting site at Daventry. Five months later, on 17 February 1935, a new transmitter opened at Droitwich radiating the Regional Programme on medium wave with a power of 50 kilowatts. At this point the original 25-kilowatt transmitter at Daventry was phased out.

Eventually, further high-power 'Twin Wave' stations would be built at Moorside Edge, Washford Cross and Westerglen. Moorside Edge, located on the Pennines, replaced the main station at Manchester and the relays at Bradford, Hull, Leeds, Liverpool, Sheffield and Stoke on 17 May 1931. Westerglen in Central Scotland, which replaced Glasgow, Edinburgh and Dundee, followed on 12 June 1932. The following year, the Washford Cross transmitter in Somerset was opened, substituting Cardiff and Swansea and also covering much of Western England.

In 1936 and 1937 additional high-power transmitting stations were established at Lisnagarvey, Burghead and Clevedon to bring the Regional Programme to most areas of the UK. Lisnagarvey covered Northern Ireland, while Stagshaw transmitted to the North East of England and Cumbria. Burghead provided cover for Northern Scotland, with low-powered transmitters at Penmon and Redmoss in Aberdeen relaying the Welsh and Scottish regional programmes. In June 1939 a transmitter at Start Point in Devon and another at Clevedon, near Bristol, enabled the last of the original transmitting stations to be closed.

The BBC's studios in Linenhall Street in Belfast were progressively enhanced and regional programming got properly underway in the province with the opening of the new transmitter at Lisnagarvey in 1936. It brought BBC radio services to audiences across Northern Ireland for the first time and was accompanied by a specially commissioned series of programmes and features.

John Reith had been eager to provide an overseas radio service since 1924 and finally, after technical and financial delays, a licence to broadcast on shortwave was acquired from the Post Office in 1926. The trial station, G5SW, opened at Chelmsford in November 1927. It was intended that G5SW would transmit programmes from Britain to the Empire from a 10-kilowatt transmitter.

How shortwave works

Shortwave signals can travel thousands of kilometres across international boundaries by bouncing off the turbulent gases of the ionosphere, the layers of electrified gas high above the earth, but it's a signal that can be somewhat capricious and subject to interference from electrical storms and other atmospheric disturbances.

The G5SW shortwave transmissions were deemed to be a success and this led to the establishment of a permanent Empire Station at Daventry in December 1932 using two 15-kilowatt transmitters and a number of directional aerial arrays to beam the signals to various parts of the globe.

The Empire Service was aimed principally at English speakers in the colonies of the British Empire, or as George V put it in the first-ever Royal Christmas Message, the 'men and women, so cut off by the snow, the desert, or the sea, that only voices out of the air can reach them.' The king's speech had been scripted by none other than Rudyard Kipling.

The Empire Service would expand in 1938, with the first foreign language service in Arabic. A European service was started in the run-up to the Second World War and French, German, Italian, Portuguese and Spanish were added.

The BBC took full advantage of emerging technology. Before 1930 the corporation had no viable means of recording sound. The first recording machine the BBC used was the Blattnerphone, which took its name from its inventor, early British filmmaker Louis Blattner. The Blattnerphone used 6-mm steel tape to record a very basic audio signal that was good enough for voice recording but not for music. The machine's spools were large and heavy, and editing was achieved by soldering the tape. The machine ran the tapes at a speed of 1.5m per second, which meant it was hazardous for the operator: a break in the tape could result in razor-edged steel flying around the studio.

There had been developments in microphone technology too. The AXBT was the fourth generation design of the original Marconi Type 'A' microphone. The ribbon microphone was particularly good in studio situations and the double-sided design, which accepted sound from front and back but not from the side, was particularly suited to voice.

It also gave the microphone its characteristic shape, which has entered popular culture as a symbolic image of broadcasting.

By the end of the 1920s the premises at Savoy Hill were becoming inadequate for the BBC's needs. The heavy soundproofing drapes and lack of ventilation made it uncomfortable to broadcast there and a few performers fainted because of the heat.

The corporation had to find and establish a new operating centre. Initially a search was made for an existing building which could be adapted for the needs of broadcasting. After looking at several properties, including Dorchester House, which would subsequently be rebuilt as a hotel, it was decided that the problems of adaptation were too great and that it would be better, despite the greater expense, to construct a new building.

Various central London sites were considered and preliminary plans were prepared for some of them. Early in 1928 a site at the corner of Portland Place and Langham Street was proposed. It was in the hands of a syndicate who were initially planning to build high-class residential flats. They offered to erect a building to suit the BBC's requirements, and to grant a long-term lease with an option to buy. An agreement between the corporation and the syndicate was signed on 21 November 1928. Incidentally, the original lease barred certain tradesmen from the site, including slaughter men, sugar-bakers and brothel keepers.

George Val Myer designed the famous and now iconic Broadcasting House in collaboration with the BBC's civil engineer, M.T. Tudsbery. The interiors are the work of architect Raymond McGrath, who supervised a team that included Serge Chermayeff and Wells Coates. They designed the studio theatre, the associated green and dressing rooms, and the dance and chamber music studios in a flowing Art Deco style.

The architects faced the classic problem of a radio building: shielding the studios from extraneous noise. They decided to place all the offices in the form of an outer shell surrounding an inner core containing the studios. The offices consequently provided sound insulation for the studios. The outer shell was constructed around a steel structure but there was still the problem of internal sound insulation. Steel would transmit sound from studio to studio so the central core was planned as a totally separate building within the outer shell. The central section was constructed almost entirely of brick.

At the front of the building are statues designed by Eric Gill. These illustrate Prospero and Ariel from Shakespeare's *The Tempest*: a fitting choice since Prospero was a magician, and Ariel a spirit of the air, in which radio waves travel.

Construction was completed in 1931 and programmes transferred gradually to the new building. The first musical programme to be broadcast from Broadcasting House featured Henry Hall and his Dance Orchestra on 15 March 1932. Stuart Hibberd read the first news bulletin on March 18. The last transmission from Savoy Hill was on 14 May 1932.

The total cost of building Broadcasting House was £500,000 and it was originally rented to the BBC at £45,000 per annum. It was later purchased for £650,000 and the freehold transferred to the BBC on 16 July 1936.

Despite all the planning, there were soon complaints about the facilities. The requirements of broadcasting had changed and expanded in the time it had taken to plan and build. Some studios were considered too small, doors were too narrow to allow equipment through and there was inadequate lighting to read scripts and scores.

Even with the efforts made by the architects, there was noise leakage between studios. The concert hall organ could be heard in other studios and conversely big bands playing in the sub-basement could be heard in the concert hall. The rumble of underground trains could also be heard in the lower studios.

The total number of BBC employees doubled in the period between 1932 and 1936 and the Empire Service made demands on space and recording facilities. St George's Hall, situated next door to the Queen's Hall, was acquired in 1933. It opened as a studio on 25 November 1933 and was mainly used for music and variety shows. The BBC installed the original BBC Theatre Organ in 1936, a Compton Melotone and Electrostatic Organ. This enabled a wide range of sounds to be produced during performances. Reginald Foort was appointed resident organist.

A space suitable for large orchestras was required so an old ice rink in Maida Vale was converted into a seven-studio complex in 1934. Over a period of fifteen months a team of one hundred men reduced the skating rink to a shell and then proceeded to rebuild it. The outside structure of the building and the arches at the doorway were preserved. The BBC Symphony Orchestra moved here when the building was reopened and it has remained their base ever since.

Although the Maida Vale building has a drama studio, it's been chiefly used for recording musical sessions. Many famous musicians have recorded there, from big bands to the Beatles. During the Second World War, Maida Vale served as the centre of BBC news operations throughout Europe. Much later Maida Vale became the home of the celebrated Radiophonic Workshop, which produced experimental electronic music for radio and television.

Sheila Barrett, the first female radio announcer, made her debut on 28 July 1933. Variety, or light entertainment as it was called at the BBC, was immensely popular in the thirties. Nevertheless, strict rules were introduced to ensure that there were no jokes about religion, drunkenness and many other sensitive subjects. Detailed guidelines were given to artists and producers and any act that breached them would have been severely reprimanded.

The Band Waggon effect

The most popular show of the 1930s was *Band Waggon*. The show only ran for a total of fifty-five episodes, but it had an enormous impact. Its memorable signature tune, extravagant musical items and irreverent humour made it pre-war radio's biggest success. *Band Waggon*, as the name might suggest, hadn't actually been devised as a comedy programme. It was to be another danceband show but the BBC decided to broaden the show's appeal by adding a resident comedian and compère.

Richard Murdoch was chosen straight away as the compère. He was a song and dance man with a background in musical revue. The choice of resident comedian was more problematic. The producers couldn't choose between Tommy Trinder and Arthur Askey so a coin was tossed to decide which one of them should get the job: heads for Trinder, tails for Askey. Heads won but as it turned out Tommy Trinder was unavailable, so Arthur Askey was invited to join the show.

The first three shows were broadcast starting on 5 January 1938, but they were a disaster. The scriptwriter was fired and a scriptwriting team of Gordon Crier, Vernon Harris, Arthur Askey and Richard Murdoch took over. Arthur Askey and Richard Murdoch soon came to dominate the show and the musical parts were reduced.

The events and characters largely existed in the minds of the listeners. Many of their sketches had Arthur and Richard sharing a top-floor flat in Broadcasting House along with Lewis the goat, and pigeons named Basil, Lucy, Ronald and Sarah. Other regular characters were Nausea Bagwash (Arthur's fictitious girlfriend) and her mother. These characters were often referred to but never heard.

The duo's adventures often ended violently with the famous *Band Waggon* crash. A corner of the stage was roped off and a large stack of assorted objects were piled up inside. At the appropriate moment it was pushed over by a sound effects man.

By the third series, Arthur Askey was in great demand for films and stage shows and Richard Murdoch had joined the RAF, therefore it was decided to curtail the series after eleven shows. The last show was broadcast on 2 December 1939 and Arthur and Richard departed from their famous flat for the final time.

When it came to major historic occasions, John Reith was no behind-the-scenes bureaucrat; he preferred to be in front of the microphone. Upon the death of King George V in 1936, Reith insisted that he make the announcement personally. The subsequent state funeral was an important milestone for the BBC and the first to be broadcast live.

Later that year, the abdication crisis gripped Britain as King Edward VIII renounced the throne to marry the American divorcee Wallis Simpson. The king's abdication speech was broadcast from Windsor Castle on 10 December 1936. Sir John Reith introduced the broadcast after giving Edward a voice test. Reith asked the outgoing monarch to read some text from the sports pages of a newspaper. Reith's diaries reveal that, during the handover, Edward bumped the table at which he was seated. An audible thump was heard and this led some listeners to think the moralistic Reith had left the room and slammed the door as a mark of disapproval.

By 1937 the BBC's headquarters in Northern Ireland was too small so a larger site was sought in the centre of Belfast. The organisation intended to build a 'Broadcasting House' similar in scope to their headquarters in London. They chose a location in Ormeau Avenue and it was estimated that the work would cost around £250,000.

Substantive building work began in 1939 and continued despite the outbreak of the Second World War. It was finally completed in 1941 at a time when regional broadcasting was largely in abeyance.

The beginning of television and the end of Reith

The BBC looked to expand its services and move into television. John Logie Baird, a Scottish engineer and inventor, had been developing the world's first working television system since 1923. In 1932 the BBC began experimental television transmissions using the 30-line Baird system. These experiments emanated from studio BB in the basement of Broadcasting House.

In June 1935 the BBC installed a mast, studio and television transmitter at Alexandra Palace and continued experiments using two systems. John Logie Baird's electromechanical system had developed into a 240-line system at this point. It was alternated with an electronic scanning system developed jointly by EMI and Marconi. The Baird system, although similar in picture quality, proved troublesome for actors, directors, and other staff to use in the studio.

The world's first regular television service started on 2 November 1936. The opening show was called *Variety* and it's estimated that only 400 'lookers in' were able to see it. The musical comedy star Adele Dixon sang a song called simply *Television*, known universally today as *The Television Song.*

Many ambitious outside broadcasts were made, including King George VI's coronation on 12 May 1937. About 10,000 people saw this on television. Other important landmarks included the first Wimbledon coverage in June 1937 and the FA Cup Final in 1938.

The Television Advisory Committee eventually recommended that the BBC adopt the 405-line Marconi-EMI system in January 1937. Viewers were treated to a wide variety of programmes, including plays, newsreels, concerts, opera, ballet, cabaret and children's cartoons. The first three television presenters were Leslie Mitchell, Elizabeth Cowell and Jasmine Bligh. Leslie Mitchell was a former actor whose handsome looks, smart suits and urbane public-school manner endeared him to a nation. Elizabeth Cowell was tall, elegant and artistic. Jasmine Bligh was attractive, humorous and could improvise easily. It's been alleged that the cameramen would put gauze over their camera lenses to soften the beautiful Miss Bligh's looks if she had been to a party the previous night.

Pay scale

Pre-war television actors were paid considerably less than their radio counterparts. The BBC gave two reasons for this. The first reason seemed perfectly valid as television had a smaller audience. However, the second was quite bizarre: the BBC reasoned that the actors should receive a reduced payment because they were shown in miniature.

John Reith disliked television intensely and was not even present at the public opening of the service in 1936. By the mid-1930s Reith's authoritarian style had many detractors. In an age of dictators it was easy to make comparisons. When Reith took his mother to tea at the House of Commons, the MP Nancy Astor asked if he got his 'Mussolini traits' from her.

Reith's own remarks didn't help. He said he respected Mussolini and talked in 1939 of Hitler's 'magnificent efficiency'. He also had a high regard for the German broadcasters: 'Germany has banned hot jazz and I'm sorry that we should be behind in dealing with this filthy product of modernity.'

Throughout this period Reith was subjected to a media campaign criticising his managerial style. In 1934 he appeared in front of the 1922 Committee of Conservative MPs, but he was able to present a petition signed by 800 BBC employees affirming their support. For a man accused of being aloof from his workforce it may be surprising to learn that he regularly took part in theatre productions by the staff amateur dramatic company. His last performance was in 1936 playing a butler in a play performed at the Fortune Theatre in London.

John Reith left the corporation on 30 June 1938. He exited Broadcasting House for the last time without ceremony, but he did carry out one more duty. Along with his secretary and his deputy, Cecil Graves, he drove to the BBC transmitter at Droitwich in Worcestershire to switch it off at midnight. He signed the visitor's book 'J.C.W. Reith, late BBC'.

There was a bitter postscript to Reith's departure. A week later the BBC governors' realised Reith had not handed in a formal written resignation and the chairman had to prompt him to do so. In a terse reply, Reith responded: 'I resign, I shall resign, I have resigned. There it is – in all three tenses…'

By the time the Second World War came, the BBC's five services (National, Regional, Overseas, Empire and Television) formed the nucleus of the greatest broadcasting company the world had ever seen.

Chapter 4

Radio in America

On 15 April 1919 the American government ended the wartime ban on public reception of radio signals. This, together with improvements in valve equipment, encouraged a number of private enterprises to be set up. The Chicago Radio Laboratory developed a 'Jeweller's Time Receiving Set'. This set received time signals transmitted by a number of government naval radio stations.

In early 1919 the USS *George Washington* had a valve transmitter installed for a transatlantic voyage. This was ostensibly to test long-range radiotelephony but they still found time to broadcast occasional concerts. One of the ship's passengers was the US president, Woodrow Wilson. It was announced that the president's Independence Day speech would be broadcast from aboard ship, but as it turned out Wilson's speech went unheard because he stood too far from the microphone.

On 30 May 1920 the Navy transmitted live from the field of an army-navy baseball game at Annapolis, Maryland. High-powered radiotelegraph stations then relayed the game worldwide. 8MT, an amateur station in Pennsylvania, broadcast advance information regarding the Uniontown Speedway races. 1DF, an amateur station in Winchester, Massachusetts, transmitted concerts on weekday nights and Sunday afternoons.

The first American news bulletin was broadcast on 31 August 1920 by station 8MK in Detroit, Michigan. It was licensed to a teenager, Michael DeLisle Lyons, and financed by Edward Willis Scripps, a newspaper magnate. The station broadcast from the premises of the *Detroit News* newspaper, and identified itself on air as the *Detroit News Radiophone*. That evening's debut featured Malcolm Bingay, the managing director of the *Detroit News*, reporting on a primary election held that day. Lois Johnson performed several songs throughout the evening.

The first college radio station began broadcasting on 14 October 1920 from Union College, Schenectady, New York, using the call sign 2ADD.

In their first month 2ADD aired what is believed to be the first public entertainment broadcast in the United States. This was a series of Thursday night concerts that were initially heard within a 160-km radius.

There was a cluster of other stations around the country. 2XG in New York broadcast a report from a football game on 18 November 1919. The station was also offering a nightly news broadcast. Another New York station, 2XX, frequently broadcast entertainment programmes from Broadway. Then there was 6XC, the 'California Theatre station', which started in April 1920.

These initial stations led to a broadcasting boom that swept across the US in early 1922. A lot of these stations were run by enthusiastic amateurs but the Department of Commerce became worried about the quality of the broadcasts. They introduced regulations that restricted public broadcasting to stations that met the criteria of a newly created broadcast service classification.

Despite these new regulations, radio stations continued to proliferate and by the end of the year there were over 500 of them transmitting to cities and towns across the country. The tremendous growth of radio broadcasting saw the development of a wide variety of innovative programme offerings. For instance, *The Man in the Moon* treated children listening to WJZ in Newark, New Jersey, to evening readings.

Not everyone was happy with the programme offerings. Writer Charles E. Duffie criticised 'the indiscriminate competitive jumble of phonograph music, uninteresting lectures, and disguised advertising talks, which have, in part, made up many programmes.' He suggested the federal government could provide a better selection of programming.

The broadcasting boom triggered a huge increase in radio-related literature. Numerous books and articles were published to introduce this exciting new innovation to the general public. There was a considerable amount of books aimed at the younger reader, many of which featured radio superficially as a prop or plot device. One notable exception to this was the Allen Chapman *Radio Boys* books. These provided the usual elements of children's fiction but they also contained comprehensive and accurate technical details.

In America, even after the rise of radio broadcasting, a few experimenters continued to try to develop another way to arrange multi-programme audio services. Early radio experimenters found that multiple low-power transmissions could be carried along telegraph, telephone or electrical wires to distant points. In 1923 there was an early (and ultimately unsuccessful) attempt to create a subscription-based 'wired radio' service in New York City. Over succeeding decades the basic idea has been developed into a wide variety of innovations, from audio services to Cable TV.

American Marconi dominated the post-war radio and communications industry. However, despite the name, the vast majority of the company's stakeholding remained in European hands. The US government wanted to avoid foreign control of its international communications so applied extensive pressure on the company to sell its operations to an American firm. American Marconi's assets were subsequently sold to General Electric, which used them to form the patriotically named Radio Corporation of America.

On its formation, the Radio Corporation of America instantly became the dominant US radio firm. Trade advertisements proclaimed that RCA was 'an all-American concern holding the premier position in the radio field'. The new company planned to build a showcase international facility, Radio Central, at Rocky Point in Long Island. The plans included ten alternator-transmitters surrounded by twelve huge antennas arrayed in spokes, each approximately 2km long. Only about twenty per cent of the planned alternator facilities were ever built, because within just a couple of years the development of far more efficient shortwave transmissions made the longwave alternator-transmitters obsolete.

RCA quickly moved into the developing broadcasting field. Its debut broadcast was transmitted on 2 July 1921 with a heavyweight boxing championship match between Jack Dempsey and Georges Carpentier. The bout was broadcast by WJY, with a transcript of the fight commentary telegraphed to KDKA in Pittsburgh, for rebroadcast by that station. There was a distinct lack of radio receivers so the majority of listeners were in halls, where volunteer amateurs set up radio receivers. Each listener was charged an admission fee, which was donated to charity.

The Westinghouse Electric and Manufacturing Company, based in Pennsylvania, also became one of the radio industry's most prominent leaders. Westinghouse was a major manufacturer of electrical appliances for the home and was the first company to widely market radio receivers to the general public.

On 2 November 1920 they began a public broadcasting service designed to promote the sale of radio receivers. The inaugural broadcast featured election returns broadcast from the company's new East Pittsburgh station. For the first few days the East Pittsburgh broadcasts went out under the special amateur call sign of 8ZZ, after which it switched to KDKA. The new station began daily broadcasts of varied offerings, which proved increasingly popular. Westinghouse quickly created three additional stations: WJZ in Newark, WBZ in Massachusetts and KYW in Chicago.

The creation of the large American radio networks was made possible by experiments undertaken by the American Telephone and Telegraph Company (AT&T). The introduction of valve or vacuum-tube amplification for telephone lines allowed them to experiment with sending speeches to distant audiences that listened over loudspeakers. These short-range experiments soon progressed to transcontinental proportions. On 11 November 1921 audiences in New York City's Madison Square Garden and San Francisco's Civic Auditorium simultaneously heard President Harding's Armistice Day speech at the National Cemetery in Arlington, Virginia.

AT&T's next step was to use the lines to establish the first nationwide network of connected radio stations. They formally announced this on 11 February 1922, although early statements referred to the set-up as a 'chain' of stations, rather than a network. AT&T's plan was to introduce what they called 'toll broadcasting'. They intended to sell the airtime to interested parties or organisations. Advertising would support the resulting programmes. Initially AT&T found it very difficult to persuade customers to purchase radio airtime. Their first success was transmitted from WEAF on 28 August 1922. It was a fifteen-minute talk promoting a Queensboro Corporation apartment complex. It cost the advertiser $50 and recouped $27,000 in sales

Although that talk has often been called 'the first-ever radio commercial', there is evidence that other stations had previously sold airtime to commercial buyers. In Jersey City Frank V. Bremer reportedly leased his amateur station, 2IA, to two local newspapers for a series of broadcasts.

The station 1XE, which was the American Radio & Research Corporation's (AMRAD) experimental station, initially broadcast from Tufts University in Massachusetts. It allegedly received money for reading stories from the *Little Folks* magazine and *Youth's Companion*. Later, when AMRAD got a new licence and started to broadcast with the call sign WGI, they started to sell airtime officially. They hired a salesman to sell thirty hours of programming a week at the rate of $1 per minute. Their first sponsored programme was by the Packard Motor Company of Boston. However, WGI's commercial operations were almost immediately suspended. It was uncertain whether this was due to the intervention of the local District Radio Inspector, or AT&T enforcing what it felt was an infringement of its patent rights.

AT&T originally thought its patent rights would give it a near-monopoly of US broadcasting and that only they had the exclusive right to sell advertising over the airwaves. By the beginning of 1923 there were over 500 broadcasting stations in America and the phone company claimed that the vast majority of these were infringing its copyrights. At this point the phone company resigned itself to the situation. They declared that it would, for an appropriate fee, licence other stations to carry on-air advertising. However, the hundreds of stations didn't rush to buy a licence.

In 1924 AT&T filed a patent-infringement lawsuit against WHN in New York City, which was eventually settled out of court. At the time of this settlement, WHN management loudly complained about AT&T's supposed plan to 'monopolise' radio. They claimed that AT&T's licence demands were stifling the growth of commercial broadcasts, though actually all stations settling with the phone company were permitted to sell advertising. They also acquired access to AT&T's lines for remote broadcasts. At this point the rest of the broadcasting stations followed WHN's lead and those that wanted to remain on the air dutifully paid for AT&T patent licences.

By the mid-1920s many broadcasting stations found themselves facing increasing financial strain. Stations were forced to buy better, but subsequently more expensive, equipment to adhere to government engineering standards. Music publishers sought royalty payments for all copyrighted music that was aired and entertainers started to demand payment for their performances. In addition to those payments there were the AT&T licence fees to consider. This led to more and more stations selling airtime. This funding system, private stations supported by on-air advertising, remains the most common method used in the United States to this day.

This willingness to let anybody hire airtime led to some strange broadcasts. WLTH in Brooklyn had a rather odd sponsor; a mysterious, bearded old man who bought a minute of time daily to declare his love for someone. During his minute of airtime he would constantly repeat the phrase 'I love you!' over and over again. He never revealed who his mysterious lover was, but as long as he paid the money the station didn't care.

After a couple of years AT&T's 'toll broadcasting' experiment finally began to produce substantial returns. Weekly network programmes such as *The Ever Ready Hour* greatly expanded advertiser interest. So the phone company pressed ahead with their national network.

AT&T originally intended to own all of the stations in the network, but ultimately they didn't need to do this. During the broadcast boom of 1922, hundreds of companies and individuals went ahead and built broadcast stations of their own. The phone company would only need to build their own broadcasting stations in two cities: WBAY and WEAF in New York and WCAP in Washington DC. Most of AT&T's network broadcasts originated from WEAF in New York City and so the network was generally called the 'WEAF Chain'.

This chain of stations was informally known as 'the Red Network'. There are a couple of stories suggesting how the network received this appellation. It's said that company circuit charts marked the inter-city telephone links in red pencil. Another story states that the colour designations originated from the red pins early engineers used to mark affiliates of WEAF. WJZ stations were marked with blue pins so they later became known as 'the Blue Network'.

The three companies that comprised the 'Radio Group' – General Electric, Westinghouse and the Radio Corporation of America – responded by creating their own, smaller, network. This centred on WJZ in New York City. AT&T blocked the rival network from using telephone lines so they had to find some other way to link up stations. At first the Radio Group used leased telegraph wires but the lines were susceptible to atmospheric and other electrical interference. Then they tried to connect the stations using shortwave radio links, but this also fell short of sound quality requirements.

In May 1926 AT&T decided that it no longer wanted to run a radio network and transferred its network operations into a wholly owned subsidiary, the Broadcasting Company of America. Then, rather unexpectedly, the Broadcasting Company of America was sold to the Radio Group companies for $1 million. They immediately shut WCAP in Washington and merged its facilities with surviving station WRC.

At this point a new company was formed, the National Broadcasting Company. This new organisation took over the Broadcasting Company of America assets and merged them with the radio group's fledgling network operations. The new division was divided in ownership between RCA (fifty per cent), General Electric (thirty per cent) and Westinghouse (twenty per cent). NBC was officially launched on 15 November 1926.

AT&T's original WEAF Chain was officially renamed the NBC-Red network and the small network that the radio group had organised became the NBC-Blue network. The Red and Blue Networks shared facilities and staff and would occasionally broadcast the same events.

To begin with, the two NBC networks didn't have distinct identities or 'formats' but evolved organically from the stations within its organisation. The Red Network, with WEAF as its flagship station, had a stronger line-up of affiliated stations and so carried the more popular, 'big budget' sponsored shows and music programmes.

The Blue Network, with WJZ as its flagship station, carried a somewhat smaller line-up of often lower-powered stations and sold airtime to advertisers at a lower cost. It mostly carried sustaining or non-sponsored broadcasts such as news and cultural programmes. The common perception of the Blue Network, even by NBC itself, was that it was a highbrow and public affairs centred network. It was the original home of the NBC Symphony Orchestra.

However, NBC encountered a certain level of criticism over its attitude to the Blue Network. Many affiliates believed the company favoured the Red Network over the Blue as they had been denied access to the World Series Baseball broadcasts. NBC also used the Blue Network as a testing ground for new talent. They frequently moved successful programmes from personalities such as Bob Hope and Jack Benny over to larger audiences on the Red Network.

On 5 April 1927 NBC reached the West Coast with the launch of the NBC Orange Network, which had KGO in San Francisco as its flagship station. The Orange Network had its own staff and facilities that, in addition to producing original West Coast programmes, also recreated productions of many eastern shows. In December 1928 a single broadcast-quality line to San Francisco was installed. This meant the Gold Network could carry eastern programming directly, but only one programme at a time.

NBC Red then extended its reach into the Midwest by acquiring two 50,000–watt clear-channel signals: Cleveland station WTAM and Chicago station WMAQ. The NBC Gold Network made its debut on 18 October 1931 and carried programmes from the Blue Network.

In 1936 a second broadcast-quality circuit was completed, this time to Los Angeles. This circuit also allowed the direction of amplification to be reversed in under fifteen seconds, allowing Los Angeles to feed broadcast-quality sound to the eastern networks as well. This meant the need for the Orange Network disappeared and it was quickly dropped. Affiliate stations were subsequently incorporated into either the Blue or Red Networks. NBC also developed a network for shortwave radio stations in the 1930s called the NBC White Network.

In a significant move in 1931, RCA signed important contracts that resulted in it becoming the lead tenant of what was to become its corporate headquarters, the RCA Building at 30 Rockefeller Plaza.

Prior to 1927 radio was regulated by the United States Department of Commerce. Herbert Hoover, who was then commerce secretary, was highly influential in the shaping of American radio. Despite his important position, Hoover's powers were limited and he couldn't refuse a broadcasting licence to anyone who requested one. This resulted in too many stations trying to be heard on too few frequencies. After several failed attempts to rectify this situation, Congress finally passed the Radio Act of 1927. This legislation transferred the administration of radio to a newly created Federal Radio Commission (FRC), although some technical tasks remained the responsibility of the Department of Commerce's Radio Division.

The Federal Radio Commission was given the power to grant and deny licences. They also allocated frequencies and power levels for each licensee. The FRC was not given any official power of censorship, although programming couldn't include 'obscene, indecent, or profane language'. Still the FRC's ability to revoke a broadcaster's licence or issue a fine obviously gave them a certain degree of control.

Many critics saw broadcasting regulation as an infringement of the First Amendment to the United States Constitution stating that the government shall not stop freedom of speech in the media. The FRC was extremely fastidious in its attempts to quash vulgar language, non-mainstream political views and 'fringe' religions.

Almost immediately, the commission was accused of bias towards large commercial radio broadcasters at the expense of smaller non-commercial broadcasters. In reality, the FRC had virtually no control of the radio networks that were in the process of dominating US radio. The networks were hardly mentioned in the Radio Act of 1927. The only reference was vague: the Commission shall 'have the authority to make special regulations applicable to stations engaged in chain broadcasting.' The act didn't permit the Federal Radio Commission to introduce any rules regulating advertising; it merely required advertisers to be identified clearly within the broadcast.

Early in 1928 the commissioners instigated a radical reorganisation of available frequencies. They forced 164 stations to justify their existence or they would be forced to stop broadcasting. Many low-powered independent stations were eliminated at this time, although eighty-one stations did survive, albeit with reduced power.

KFKB in Kansas was one of the most popular stations in America. The station's owner was a controversial medical doctor called John R. Brinkley. He repeatedly advocated the use of his own 'alternative' medications on air such as the transplantation of goat glands into humans to cure male impotence. In 1930 the Federal Radio Commission denied his request for renewal. Undeterred, he simply beamed his programmes to the United States from the powerful Mexican station XER.

XER was the first of many Mexican stations to aim its programmes northwards. These stations, known as 'border blasters', could be heard over large areas of the US. They were also referred to as 'X Stations' for their call letters: Mexican stations are assigned call signs beginning with XE or XH, whereas American stations begin with the letters W or K. Canadian stations begin with C or VO.

The Mexican stations didn't have to adhere to US broadcast laws and often passed themselves off as American. They would even disguise the fact by using a Texas post office box as a mailing address. The more unscrupulous stations, unfettered by the constraints of US regulations, were free to peddle dubious goods. One such station hawked prayer handkerchiefs and claimed these items negated the need for a doctor as they would cure any ailment. Another radio evangelist reportedly was trying to sell listeners autographed photos of Jesus Christ.

During the 1930s radio technology had developed to such an extent that simultaneous global broadcasting became possible. In June 1930 General

Above left: Heinrich Hertz, the discoverer of radio waves, pictured c.1890. (Wikimedia Commons)

Above right: A portrait of Sir Oliver Lodge c.1910. His inventions made it possible to tune a radio to a desired frequency. (Library of Congress, Prints and Photographs Division, Washington, D.C. 20540 USA)

Above: Edouard Branly. Discoverer of the coherer c.1911. (Wikimedia Commons)

Right: Karl Ferdinand Braun c.1909. (Wikimedia Commons)

Left: Jagadish Chandra Bose, Physicist at Calcutta University. (Wikimedia Commons)

Below: Charles Herrold and his assistant Ray Newby c1910. (Wikimedia Commons)

Charles Herrold (standing in the doorway) in his laboratory c.1912. (Photograph included in the article *Experiments on Ground Antenna with Their Relation to Atmospherics* by Charles D. Herrold, Page 11 of the July 1919 issue of *Radio Amateur News*)

Reginald Fessenden's alternator transmitter at Brandt Rock, Massachusetts. (Uncredited magazine article illustration - Illustration on page 49 of the 26 January 1907 issue of *The American Telephone Journal* magazine, included in the article *Experiments and Results in Wireless Telephony* by John Grant)

Above left: A publicity photograph of Guglielmo Marconi posing with his early radio apparatus. (Unidentified photographer. Smithsonian Institution from United States)

Above right: Marconi's Poldhu station c.1900. This photograph shows one of his early monopole antennas. The word antenna, which means 'tent pole' in Italian, originated from Marconi. (Wikimedia Commons)

The first antenna system at Poldhu. The antenna collapsed during a storm on 17 September 1901. (Wikimedia Commons)

Above: The second antenna system at Poldhu. This photograph shows the antenna used in the first transatlantic radio transmission on 12 December 1901. (Wikimedia Commons)

Below: Marconi's station at Connemara. (Photographer: Robert French Collection: Lawrence Photograph Collection)

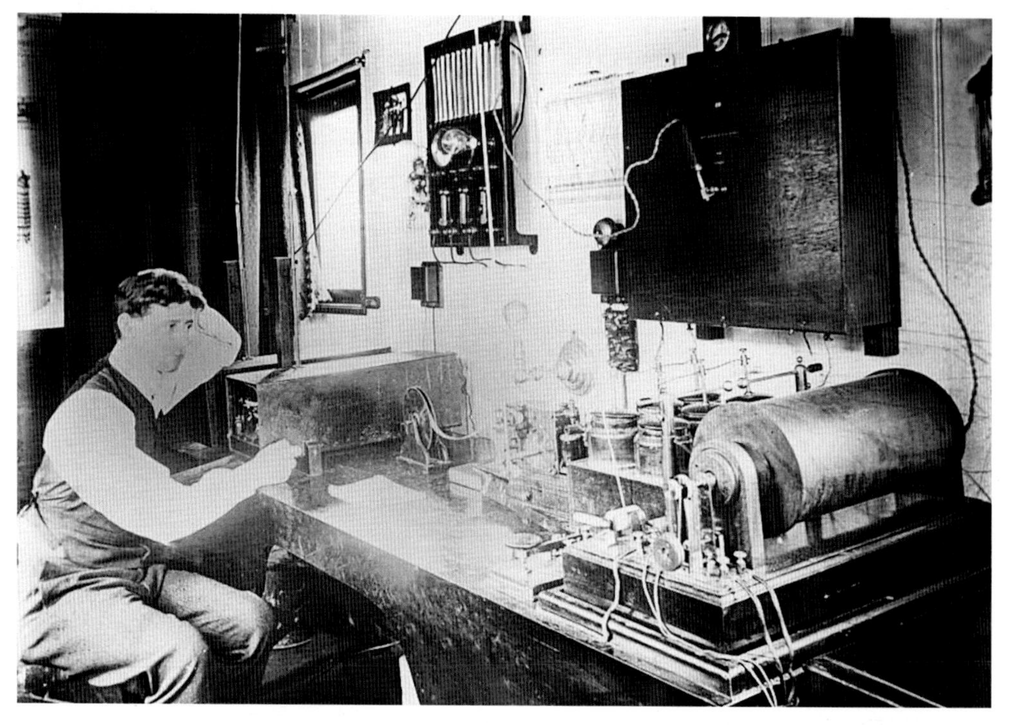

A Marconi operator at work on board the ship *Deutschland.* (Library of Congress. George Grantham Bain Collection)

Above: A Marconi Type 106 crystal radio receiver. (*Practical Wireless Telegraphy*, Revised Ed., Wireless Press Inc., New York)

Left: A typical Marconi wireless room. (National Library of Ireland Ref.: P_WP_2539)

Right: Reginald Fessenden's wireless station at Cape Hatteras in North Carolina. (General Negative Collection, State Archives of North Carolina, Raleigh, NC)

Below: A farmer listening to a typical home crystal set. (Raymond Frances Yates, Louis Gerard Pacent 1922. *The Complete Radio Book*, The Century Co., New York)

Above left: Nathan B Stubblefield with his wireless telephone. (Nathan B. Stubblefield Collection at Murray State University Jackson Purchase Digital Archives)

Above right: Dame Nellie Melba sings at Chelmsford. (November 1920 issue of *The Wireless Age*, which was included in the article 'A Newspaper's Use of the Radio Phone')

A 'radiophone dance' held by an Atlanta social club in May 1920. The participants danced wearing earphones to music transmitted from a band across town. ('Dancing by Radiophone' in *Radio Amateur News*, Experimenter Publishing Co. Inc., New York)

Above left: Edwin Armstrong and a super regenerative receiver. (*Popular Radio*, Popular Radio Inc., New York, Vol. 2, No. 3, November 1922)

Above right: An example of a homemade superheterodyne receiver circa 1920. (Paul F. Godley, 'High Amplification at Short Wave Lengths', in the *Wireless Age*, Wireless Press, Inc., New York, Vol. 7, No. 5, February 1920)

Right: Edwin Armstrong with his portable superheterodyne receiver. (*Radio News* magazine, Experimenter Publications, Inc., New York, Vol. 5, No. 10 April 1924)

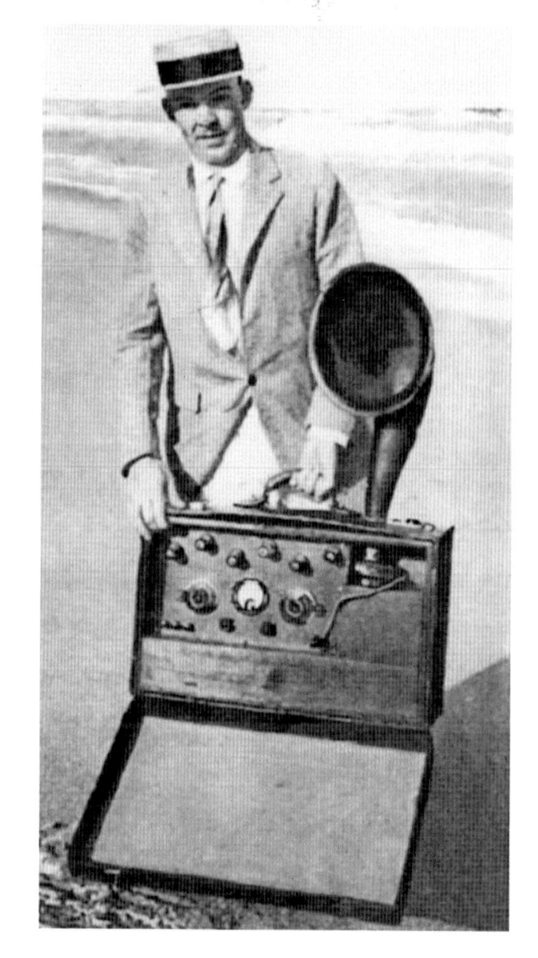

Left: PCGG in Holland's station transmitter c.1922. (*Illustrated London News*, 29 April 1922, Part of the article 'A Dutch Concert Heard in an English Home')

Below: An advertisement recommending sets to listen to PCGG's Sunday Concerts. (Advertisement from the front cover of the 12 November 1921 issue of *The Wireless World*)

DUTCH CONCERTS

Marvellous results have been secured by those using the "TINGEY" 2 and 3 VALVE RECEIVERS

1 Valve H F.—1 Rect.—1 L.F.
PRICE OF RECEIVER
(Without Valves)
So designed as to be equally efficient on all wavelengths.
£8 10 0

"Montpellier,"
Grosvenor Road,
Southampton.

Dear Sir,

The new 3-valve set and inductances I am delighted with, I get better results than I have done with a 6-valve Amplifying set

The strength of signals are *terrific*, and may tell you I had the Dutch Concert all the afternoon.

The Band Selection and Clarionet Solo were very fine also the singing was quite good.

On Thursday last, in spite of the strong atmospherics, I got the Hague music from the start to finish of the transmission, also from 2.30 p.m. yesterday (Sunday) till 6 p.m.

The Music was particularly clear, also speech which was a very lengthy one.

I may add without any exaggeration that nearly all signals are readable with head gear on table. Horsea—Coltana—Paris—Devizes—Poldhu can be easily read 30 to 40 feet away from the 'phones.

I must say that your patent Inductances are a great success, by far the best I have ever tried. More than delighted with 3-valve receiver the selectivity of tuning is very fine. Thanking you for prompt delivery.

I am, yours truly,
(Signed) C. WHITE.

The letter quoted above is the first of several encouraging letters recently received by us and is published with the kind permission of our customer.

HAVE YOU RECEIVED OUR NEW PRICE LIST ?

1 Valve H.F.—1 Rect.
PRICE OF RECEIVER
(Without Valves.)
So designed as to be equally efficient on all wavelengths.
£6 0 0

Office and Works
Telephone:
1916 Hammersmith

W. R. H. TINGEY
Specialist in Wireless

Offices and Showrooms
92, Queen Street,
Hammersmith, W.6

Above: A family gather around their set. (National Publishing Co. Library of Congress)

Right: An issue of *Radio Journal of Australia*. This short-lived magazine was published between November 1927 and March 1928. (*Radio Journal of Australia*, volume 1 number 2, 30 November 1923 in Sydney, New South Wales, Australia)

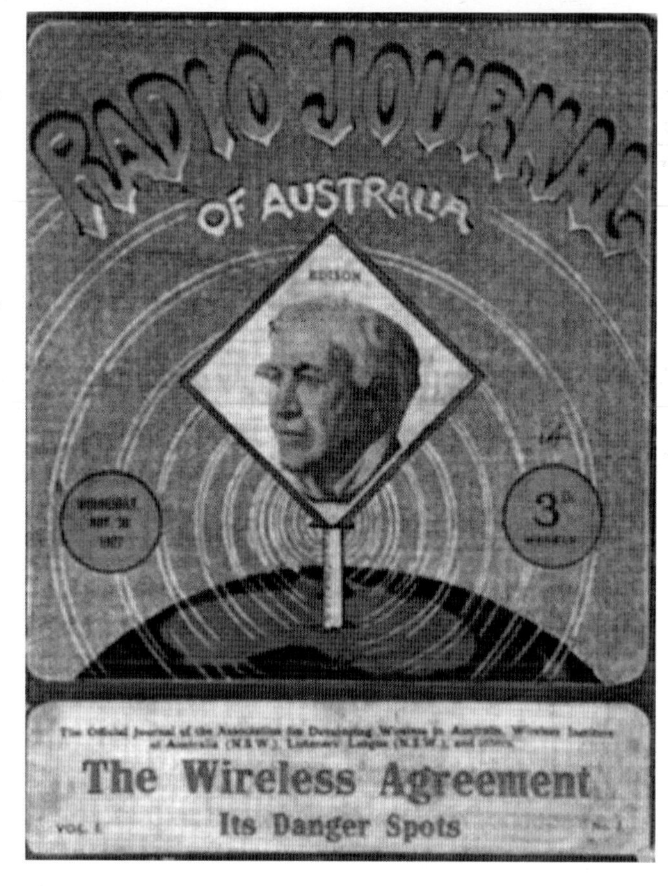

An example of a tuned radio frequency receiver from 1925. (S. Gordon Taylor, *How to get the most out of your ready- made receiver* in *Popular Radio*, published by Popular Radio Inc, New York)

A cartoon from the *Radio Times* magazine, 5 October 1923.

Above: A rehearsal in the studio of 2GZ in New South Wales, Australia. (The State Library of New South Wales)

Right: Lord Reith. The first Director General of the BBC. (National Portrait Gallery from Central Office of Information)

Above left: A Brisbane household listen to the radio c.1942. (John Oxley Library, State Library of Queensland)

Above right: Leonard F Plugge. Owner of the International Broadcasting Company. (Wikimedia Commons)

Below: Sir Ernest Fisk inspects one of his factories. (From the collection of The State Library of New South Wales)

Above: The cast of *ITMA* visit the home fleet at Scapa Flow. (Mason H A (Lt), Royal Navy official photographer. Photograph A 21269 from the collections of the Imperial War Museums)

Right: Orson Welles as *The Shadow*. (Mutual Broadcasting System)

Above: Winston Churchill makes his VE Day broadcast from 10 Downing Street on 8 May 1945. (Photograph H 41846 from the collections of the Imperial War Museums)

Left: Bing Crosby's CBS publicity photograph. 1951. (CBS Radio)

Electric engineers beamed a popular song, *I Love You Truly*, around the world via shortwave relays and rebroadcasts. They followed that with another stunt broadcast. In New York, GE engineers twisted a cat's tail and it meowed into a microphone. A dog, placed by a loudspeaker in Australia, heard the cat and barked into a nearby microphone. Back in New York the cat responded by arching its back with its fur standing on end, thus completing the world's first transoceanic dog-and-cat fight.

As the big networks expanded, many independently owned radio stations applied to join them. The stations that carried network programming were termed as affiliates. These affiliates agreed to carry designated network programmes. Since the programmes included commercials, the stations received a share of the network revenue. At the same time, the affiliates could run their own local commercials around the network programmes.

The rise of NBC and CBS

Between 1930 and the mid-1950s NBC was the dominating force in American radio. The network's stations were frequently the most powerful in their area. Many broadcast on clear frequencies that were capable of transmitting hundreds or thousands of kilometres at night, though the main reason behind NBC's popularity had to be the programmes that were transmitted during this period. American radio's earliest mass hit was broadcast by NBC: *Amos 'n' Andy* began in 1926-27 as a fifteen-minute serial and the two struggling title characters appealed to a broad audience, especially during the Great Depression.

NBC became home to many of the most popular performers and programmes on the air: Bob Hope, Al Jolson, Jack Benny, Edgar Bergen and Fred Allen were all mainstays of the network. Popular programmes included *The Great Gildersleeve, Vic and Sade, Death Valley Days,* and *Fibber McGee and Molly.*

The famous three-note NBC chimes were first used in 1929, but evolved after several years of development. The chimes were first heard on WSB in Atlanta. Someone at NBC in New York heard the three-note sequence during the networked broadcast of a Georgia Tech football game and asked permission to use it on the national network. NBC started to use the three notes in 1931 and they were mechanised in

1932 and used as a cue for switching different stations between the Red and Blue network feeds.

The NBC chimes were the first audio trademark to be accepted by the US Patent and Trademark Office. Contrary to popular legend, the three musical notes, G-E-C, didn't stand for the General Electric Company. General Electric was forced to sell its share of RCA in 1930 because of antitrust charges. A variant sequence, known as 'the fourth chime', was used during wartime and in time of disasters.

NBC's main rival was the Columbia Broadcasting System. CBS began as United Independent Broadcasters with sixteen affiliates. When Columbia Records invested in the radio network in 1927, it became the Columbia Phonographic Broadcasting System. The Paley family, who owned a tobacco firm, secured majority ownership of the network in 1928. The name was streamlined to the Columbia Broadcasting System at this point.

Samuel Paley's intention had been to use his acquisition as nothing more than an advertising medium for promoting the family business. This proved a shrewd move as cigar sales more than doubled after a year. The new owners soon found that they could sell far more than just cigars over the airwaves. Sam Paley handed control of the radio network to his twenty-six-year old son, William S. Paley. Within a decade, under Bill Paley's inspired leadership, the CBS network expanded to 114 affiliate stations with WCAU as the flagship station.

Paley built his empire by sweet-talking the very best performers and technical talent of the era to join CBS. He brazenly stole shows and performers from competing networks and his relentless pursuit of the best talent in the industry paid dividends for his network.

CBS were at the heart of an event that became known as the press-radio war. Like their European counterparts, the American newspapers saw radio as a serious competitor. The major news services, including the Associated Press (AP), the International News Service (INS), and the United Press (UP), launched a battle against the radio stations. The agencies threatened to cut off radio's flow of news, and could do this by restricting access to Teletype machines. The Teletypes supplied the country's newspapers with regular summaries of news, feature stories, weather forecasts and bulletins.

Recognizing the consequences of this threat, CBS assembled their own news gathering agency. This enraged the newspaper groups who issued a list of demands. The first demand was the total cessation of CBS' news gathering operations. They also tried to restrict the amount of news broadcasts to two news summaries a day, which could only be aired after the morning and afternoon newspapers were published. They further stipulated that the newscasts could not be sponsored. CBS eventually won the battle and proceeded to assemble one of the most respected and knowledgeable news teams in the history of radio broadcasting.

Radio stations gradually reduced their commitments to news. They realised that news gathering was hugely expensive. Ratings also indicated that most listeners were more interested in hearing entertainment than news. Nowadays most American stations switch to an audio network on the hour for a short news summary

CBS radio produced dozens of long-running shows. One of the most fondly remembered is *Suspense*, which ran from 1942 to 1962. The programme was subtitled 'radio's outstanding theatre of thrills,' and focused on suspense, thriller-type scripts, usually featuring leading Hollywood actors of the era. Approximately 945 episodes were broadcast during its long run.

The War of the Worlds

CBS was responsible for the most infamous broadcast in American radio history. On the evening of 30 October 1938 the Mercury Theatre's adaptation of H.G. Wells' science fiction tale, *The War of the Worlds*, caused widespread mayhem. Rather than a straightforward dramatic adaptation of the novel about an alien invasion from Mars, Orson Welles chose to portray the events in the book differently. He changed the location and time from Victorian England to contemporary America. More importantly, the first half of the story was presented as a simulated broadcast with mock news bulletins and on-the-spot reports. It was undoubtedly this novel dramatic device that led to the subsequent panic.

Writer Howard Koch chose Grover's Mill as the location for the first Martian landing by randomly prodding a pencil into a map. He then plotted the Martians' route towards New York City, defeating the army and destroying dozens of familiar place names along the way.

Unfortunately, listeners tuning in late had absolutely no idea that what they were experiencing was a dramatic presentation.

In the programme makers' defence, the show was clearly advertised as such. There were also several clear announcements before, during, and after the broadcast that clearly stated that the show was a dramatisation of the H.G. Wells short story. Nevertheless, many listeners thought the broadcast was fact, not fiction. People fled their homes in terror, carried weapons in an attempt to defend themselves against aliens and even wrapped their heads in wet towels as protection from Martian poison gas. CBS stations, city authorities and police forces around the country received hundreds of telephone calls from panic-stricken listeners.

News of the panic quickly generated a national scandal. In the days following the adaptation there was widespread anger towards the perpetrators. The programme's news-bulletin format was condemned as maliciously deceptive by some newspapers. Gradually the chorus of disapproval subsided and many public figures spoke out in favour of the broadcast. In a perceptive column in the *New York Tribune*, Dorothy Thompson stated that the broadcast exposed how politicians could utilise the media to create theatrical illusions and manipulate the public.

MBS

The Mutual Broadcasting System operated from 1934 to 1999. Of the four national networks during American radio's classic era, the MBS had the largest number of affiliates. Despite this, the network was constantly in a precarious position financially. For the first eighteen years of its existence MBS was owned and operated as a cooperative. Unlike NBC and CBS, which distributed programmes to affiliated stations, the original Mutual was an arrangement among its four founding stations to share programmes produced at the stations.

Mutual's original participating stations were WOR in New Jersey, WGN in Chicago, WXYZ located in Detroit, and Cincinnati's WLW. As they were located in the two largest markets, WOR and WGN provided the bulk of the programming. WXYZ left Mutual in 1935 and was replaced by CKLW in Ontario, Canada.

The MBS struggled to make ends meet in its early years. In 1935 the cooperative only sold $1.1 million dollars in advertising. Compare that amount to the $26.6 million in sales logged by NBC and $16.3 million achieved by CBS that same year. It was only when the Colonial and Don Lee regional networks were amalgamated into MBS in 1936 that it achieved national coverage. In 1947 Mutual had 400 affiliates and that figure had risen to 560 by 1952. At this point the General Tire Company purchased the network and its days as a co-operative were over. After this the MBS was purchased by a succession of conglomerates, including Armand Hammer, The Hal Roach Studios and 3M.

MBS's shaky financial position influenced the type of programming it aired. Entertainment and music shows were expensive to produce so Mutual aired a lot of low-budget dramatic shows. Most of these were produced at WOR using a repertory company of New York actors. The *Mysterious Traveller* is a fine example of the types of show produced. Children's serials were also relatively cheap to produce and *Red Ryder* and *The Adventures of Superman* generated a lot of revenue for the network. Sport was another lucrative money earner for Mutual. For many years it broadcast commentaries on major sporting events such as the Baseball World Series and Notre Dame Football.

Despite limited resources, MBS produced some of the best programmes of the era. *Sherlock Holmes, Dick Tracy*, *The Green Hornet* and *The Shadow* were all big hits on the network. The latter programme is arguably the most fondly remembered show of the period. *The Shadow* was undoubtedly the most popular crime show on the radio, often drawing as many as fifteen million listeners.

The Shadow had been featured on the NBC and CBS network between 1930 and 1933, but this version had been a crime anthology series where the Shadow was merely the narrator. When he first appeared on the Mutual Network in 1937, *The Shadow* had been reinvented as a man of action. According to the show's introduction, the Shadow was, 'in reality, Lamont Cranston, wealthy young man about town.' Cranston had been taught a 'strange and mysterious secret, the power to cloud men's minds so that they cannot see him.' The Shadow used these powers of invisibility and hypnosis to solve crimes. Joining Cranston on his adventures was Margo Lane, his 'friend and companion'.

The episodes were largely formulaic and one of the most common themes revolved around the failures of scientific discovery. Villains were frequently scientists or intellectuals who used their research for evil rather than the benefit of mankind. Margot Lane was the stereotypical damsel in distress in most episodes, finding herself in the clutches of that week's villain. It was up to Cranston to save her without revealing his identity.

The 'Golden Age'

The 1930s through to the 1950s are often referred to as the 'Golden Age of Radio', when broadcasts caught the public imagination. In 1933, 3.6 million radio sets were sold and this was during the Depression. By 1939 about eighty per cent of Americans owned radios.

Radio borrowed much of its broadcast format from other entertainment media. During the 1930s *The Amos 'n' Andy Show* was an astounding success and harked back to an older entertainment tradition: the minstrel show. This was when white entertainers painted their faces black and pretended to be 'Negroes'. Two white comedians, Freeman Gosden and Charles Correll, brought this dubious genre to radio. Nevertheless, Amos and Andy were so popular that telephone use dropped fifty per cent during their broadcasts and cinemas interrupted screenings to pipe in the programme.

Another unlikely radio success was ventriloquist act, Edgar Bergen and Charlie McCarthy. The popularity of a ventriloquist on radio, when nobody could see neither the dummy nor his skill, mystified many critics. It was Bergen's skill as an entertainer and vocal performer, and especially his characterisation of Charlie that carried the show. Edgar and Charlie were continuously on the radio until 1956. Peter Brough paralleled Bergen's success on radio in the United Kingdom. The ventriloquist and his dummy Archie Andrews featured in the BBC show *Educating Archie.*

Music was also a principal source of programming for radio. CBS sponsored broadcasts by the New York Philharmonic and NBC formed an orchestra around the talents of Italian maestro Arturo Toscanini. Undoubtedly the most popular music on radio came from the theatres and dance halls. Singers such as Ella Fitzgerald, Rudy Vallee and Dinah Shore became firm audience favourites.

Famous swing and jazz stars such as Benny Goodman, Tommy Dorsey and Duke Ellington broadcast regularly. These transmissions were almost always live, as few records were licensed for radio broadcast. This was because of a long-standing royalty dispute between broadcasters and the American Federation of Musicians and American Society of Composers, Authors and Publishers, which effectively barred the use of recorded music on American radio.

Sports programmes became instantly popular. Baseball, boxing, and college football all succeeded in attracting audiences. The future American president Ronald Reagan was a sports announcer at WHO in Iowa during the 1930s and regularly commentated on baseball games by the Chicago Cubs. During away games Reagan remained in the Des Moines studio He was fed information from the game by telegraph and invented the remainder of the action.

The birth of the 'soap opera'

A more original staple of radio was the daily serial aimed largely at a female audience. These broadcasts became known as 'soap operas' because household-products advertisers, such as soap makers, usually sponsored them. The soaps served up daily doses of romance and melodrama to their enthusiastic audience. The first soap opera was *The Romance of Helen Trent*, which made its debut in 1933 and lasted until 1960. Other favourites included *Just Plain Bill, Our Gal Sunday* and *Ma Perkins.*

The most popular dramas tended to be crime and suspense stories, occasionally drawn from contemporary comic books, pulp magazines and classic fiction. They included *Charlie Chan, Tarzan, Buck Rogers in the Year 2430* and *Jack Armstrong, the All-American Boy*. The horror anthology series was also fashionable. *Lights Out* was a particularly gruesome example of this genre. Sound effects technicians would gleefully accentuate events happening within the drama with outlandish sounds. For example, they would produce the sound of splattering blood and guts by immersing a bathroom plunger in warm spaghetti.

At the beginning of the Golden Age, American radio network programmes were almost exclusively broadcast live. The national networks prohibited the airing of recorded programmes until the late 1940s. This was mostly due to the inferior sound quality of phonograph discs. As a result, prime-time shows would be performed twice, once for each coast. The reason why many of these shows survive today is because 'reference recordings' were made as they were being broadcast. This would have been for review by the sponsor or for the network's own archival purposes. The pre-recording of shows became more common after the Second World War, when high-fidelity magnetic wire and tape recording techniques were developed.

Local stations had always been free to use recordings and sometimes made extensive use of pre-recorded syndicated programmes. When a substantial number of copies of an electrical transcription were required, the same process used to make ordinary records was used. An electroplated master recording was cut, from which pressings in vinyl were moulded in a record press.

These recordings were made using a cutting lathe and acetate discs. Programmes were usually recorded at 331/3 rpm on 16-in discs. Sometimes, the groove was cut starting at the inside of the disc and running to the outside. This was useful when the programme to be recorded was longer than fifteen minutes so required more than one disc side. By recording the first side outside in, the second inside out, and so on, the sound quality at the disc changeover points would match and result in a more seamless playback.

An inside start also had the advantage that the thread of material cut from the disc's surface, which had to be kept out of the path of the cutting stylus, was naturally thrown toward the centre of the disc so was automatically out of the way. When cutting an outside start disc, a brush could be used to keep it out of the way by sweeping it toward the middle of the disc. Well-equipped recording lathes used the vacuum from a water aspirator to pick up the cuttings and deposit them in a water-filled bottle. In addition to convenience, this was also safer as the cellulose nitrate thread was highly flammable and a loose accumulation of it combusted violently if ignited.

Most recordings of radio broadcasts were made at a radio network's studios, or at the facilities of a network-owned or affiliated station, which might have four or more lathes. A small local station often had none. Two lathes were required to capture a programme longer than fifteen minutes without losing parts of it while discs were flipped over or changed, along with a trained technician to operate them and monitor the recording while it was being made.

In the mid-1930s American radio stations got a new regulatory body to oversee their development. In 1934 Congress passed the Communications Act, which abolished the Federal Radio Commission. The new Federal Communications Commission took over the existing responsibilities of the FRC and also acquired jurisdiction over telephone and telegraph companies.

In 1939 the FCC thought that NBC's two networks gave it an unfair advantage over its competitors and ordered RCA to divest itself of one of the two networks. RCA fought the divestiture order but divided NBC into two companies to pre-empt any judgement going against them. The Blue network became the 'NBC Blue Network, Inc.' and the NBC Red became 'NBC Red Network, Inc.' There followed a long legal battle but RCA lost its final appeal before the US Supreme Court.

On 10 January 1942 the two networks had their operations formally divorced. The Blue Network was referred to on air as either 'Blue' or 'Blue Network,' with its official corporate name being Blue Network Company, Inc. NBC Red, on the air, became known as simply NBC on 1 September 1942. In May 1943 RCA sold the Blue Network for $8 million to Edward J. Noble, who also owned Life Savers candy and the Rexall drugstore chain

Noble wanted a better name for the network and in 1944 acquired the rights to the name American Broadcasting Company. The Blue Network became ABC officially on 15 June 1945. Therefore ABC and NBC, although bitter rivals nowadays, are essentially fraternal twins separated by a custody battle.

ABC was not an instant success and had to build an audience gradually. It did this by purchasing more stations. The most significant addition was the highly profitable Detroit station WXYZ. This station was where *The Lone Ranger* and *The Green Hornet* originated, although these programmes were not included in the deal. Noble also bought KECA in Los Angeles and this gave the network a Hollywood production base. Although ABC was still in fourth place by the late 1940s, they began to gain ground on the better-established networks.

Alternative programming became an ABC speciality. They would broadcast a boisterous quiz show like *Stop the Music* against more sedate offerings on the other networks. ABC was the first to utilise advances in tape-recording technology brought back from Europe. The new network pre-recorded many programmes, as the audio quality of tape was equal to that of 'live' broadcasts. Consequently, ABC was able to attract several high-rated stars who wanted freedom from rigid schedules. Bing Crosby was probably the star with the highest profile who joined ABC during this period.

After the Second World War, Sunday night became the main ratings period with all the networks jostling for the top spot. In the late 1940s CBS gained ground by allowing radio stars to use their own production companies, which proved highly profitable for them. Jack Benny, the nation's top radio star at the time, was the first of many NBC performers to transfer to CBS.

NBC hit back with *The Big Show* in November 1950. This ninety-minute variety show updated radio's earliest musical variety style with sophisticated comedy and dramatic presentations. It regularly featured prestigious entertainers such as Louis Armstrong, Groucho Marx and Bob Hope. The show's regular hostess was stage legend Tallulah Bankhead. However, *The Big Show*'s initial success didn't last, despite critical praise, as most of its potential listeners were increasingly watching television instead. The show limped on for two years before cancellation. It's estimated that NBC lost a million dollars on the project as they were only able to sell advertising time during the middle half-hour every week.

NBC's last major radio programming push of the golden era began on 12 June 1955. *Monitor* was a continuous all-weekend mixture of music, news, interviews and features. The programme was a success for several years, but listening figures dipped during the 1960s. Local stations, especially in larger markets, were reluctant to break from their established formats to run non-conforming network programming. When *Monitor* finished its run on 26 January 1975 very little remained of NBC network radio. The only things left were hourly newscasts and news features.

The NBC News and Information Service launched on 18 June 1975. It provided up to fifty-five minutes of news per hour around the clock to local stations that wanted to adopt an all-news format. Unfortunately, the service failed to attract enough stations and was discontinued in 1977.

The end of the era now known in America as the 'Golden Age of Radio' is hard to pinpoint. Most people agree that it falls somewhere between 1957 and 1962, but there are several milestones that account for radio's gradual decline. In the late 1930s the royalty dispute between musicians and broadcasters was settled. This agreement led to more recorded music being transmitted and the end of live broadcasts of orchestras. The federal government was about to crack down on radio station networks on the basis that they violated antitrust laws. Most important of all, at the 1939 New York World's Fair, the public was introduced to television. Although the Second World War would delay the deployment of this new technology, it signalled the end of radio's media dominance.

By the middle of the 1950s a lot of radio stars had transferred to television, and the networks focused their resources on the new visual medium. Radio audiences went into a steep decline and many popular programmes were axed. The cancellation of two popular programmes on the evening of 30 September 1962 is often cited as the end of the golden age of radio in America. The last episode of *Yours Truly, Johnny Dollar* ended at 6.35pm Eastern Time, followed immediately by the final broadcast of *Suspense*. When these two radio favourites ended that gave a clear signal that the 'Golden Age' was well and truly over.

Chapter 5

Voices from Europe

In 1922 the British government awarded a monopoly broadcasting licence to a single British Broadcasting Company. This arrangement lasted until 1927, when the broadcasting licence of the original BBC was allowed to expire. The assets of the former commercial company were then sold to a new non-commercial British Broadcasting Corporation, which operated under a UK charter from the Crown. This meant there was no possibility of commercial broadcasters operating from inside the United Kingdom. The BBC undoubtedly provided quality programming of great broadcasting worth, but some listeners found this type of programming dull and monotonous and looked for an alternative.

The early British radio experimenters were able to calibrate their wireless sets by tuning in to two continental stations that transmitted time signals at regular intervals. These were a station with the call sign FL based at the Eiffel Tower in Paris and a German station POZ at Nauen. The stations became immediately recognisable by the signals that emanated from them: the French station emitted a deep low note, while the German produced a higher, usually weaker, tone. 'Nauen squeaks and the Tower squawks' became a memorable phrase for enthusiasts to enable them to distinguish between the two stations.

Experimenters would often try to hear stations from the continent. This practice was called 'searching the ether'. A contender for the earliest regular broadcast station in Europe would have to be OTL, which was based in the grounds of the Royal Castle at Laeken in Belgium. This largely forgotten radio station, run by Robert Goldschmidt, made its first broadcast on 23 March 1914. Its aim was to transmit a programme of musical entertainment for public consumption every Saturday evening at 5pm. The station received good reception reports from several hundred listeners based in Belgium and Northern France, and broadcast its concerts until German troops destroyed the transmitter when they entered Brussels in the early days of the First World War.

The station PCGG in Holland was another early favourite of radio experimenters. The station was run by Hans Idzerda to promote sales of his company's crystal sets and components. Idzerda had settled in Scheveningen as an independent consultant 'for the application of electricity in every area'. During the First World War he designed and built equipment for the Dutch government to help them monitor the movement of Zeppelins.

He also worked in conjunction with Philips, the Dutch electrical firm, to develop a triode valve that revolutionised radio reception. Philips agreed to manufacture his valves provided he would buy at least 180 of them in a year. Production began in 1918 and in the first year he sold 1,200.

The Philips IDEEZET 'soft' triode valve was named after the first three letters of Idzerda's surname. Basically, it was an extension of a vacuum diode with a third electrode in the form of a lattice, the steering grille, placed between the cathode and anode. The valves were made from components of small incandescent lamps. Its invention founded the electronic age, making possible amplified radio technology and long-distance telephony.

PCGG evolved from a 'one-off' transmission in 1919 of a programme called *Soiree Musicale*, which was transmitted from Idzerda's home in The Hague. It was transmitted between 8pm and 11pm on 6 November and utilised a radio telephone transmitter that Idzerda had designed and built himself. His colleagues at Philips sponsored the transmission and the broadcast reached an unexpectedly large audience. Idzerda was then encouraged to introduce regular transmissions.

The station was useful to listeners in gauging the efficiency of their set. If they could pick up *The Hague Sunday Afternoon Concert* on 280 kilohertz then it was a good receiver. Concerts were also transmitted on Monday and Thursday evenings. Multilingual announcements were made in Dutch, French and English.

PCGG's broadcasts became very popular and elicited a great deal of listener correspondence. Idzerda managed to secure sponsorship, some from as far afield as the United Kingdom. Unfortunately, he got into financial difficulties and his licence was withdrawn on 11 November 1924. His company was declared bankrupt a month later.

Undaunted, he created another company called NV Idzerda Radio and resumed broadcasting. Unfortunately, he was unable to secure a licence with sufficient power and was restricted to night-time broadcasts only. This made it difficult to attract sponsorship and make the enterprise financially viable. He finally capitulated in 1935 and donated what was left of his equipment to the Dutch Postal Museum.

Idzerda worked for the Dutch resistance during the Second World War. During an undercover mission at Wassenaar rocket launch site in November 1944, he was caught with remnants of a V2 rocket in his possession. He was arrested as a spy and summarily executed a few weeks later.

There were more commercial experiments throughout the 1920s. In 1925 Selfridge's in London sponsored a fifteen-minute fashion talk from a studio on the Eiffel Tower. In 1927 and 1928 the Kolster Brand radio manufacturer sponsored a string of English concerts by the De Groot Orchestra from Hilversum.

A rather unusual commercial radio venture took place in 1928. The *Daily Mail* newspaper group chartered a steam yacht called *Ceto*. They renamed it the 'Broadcasting Yacht' and it could be accurately described as an early forerunner of the 1960s 'pirate radio' ships. The idea was to broadcast at sea from just outside the three-mile limit and advertise the *Daily Mail*, the *Evening News* and the *Sunday Dispatch*.

They set off from Dundee with a small transmitter on board but soon encountered problems. They were only able to broadcast when the sea was calm and the idea of transmitting at sea was swiftly abandoned. The German firm of Siemens Halske offered a solution by supplying four powerful loudspeakers and mounting them on the yacht's superstructure. This public address system was capable of being heard clearly for more than 3km on a moderately clear day.

They were able to continue their cruise around the coast of Britain, blasting out gramophone records and publicising their newspapers. En route they would stop off at various resorts to be welcomed by civic dignitaries and 'broadcast' special concerts. The *Ceto*'s final tour of duty took her back round to London, mooring up at Tower Bridge on 1 September, by which time the crew had visited eighty-seven resorts and coastal towns and undertaken 300 broadcasts.

From 1929 to 1931 the Vocalion Record Company sponsored an occasional series of record programmes from Radio Toulouse.

They were not the only record company to realise the potential of radio. Decca Records sponsored a programme on Radio Paris with both French and English announcers, who would give the details of the record played and invite listeners' requests for the following programme.

The fashion talk aired from the Eiffel Tower in 1925 had been arranged by British entrepreneur Leonard Frank Plugge. He was a former RAF captain and engineer on the London Underground. Plugge was a radio enthusiast and listened intently to the continental stations. He quickly became obsessed with the new medium, and its commercial possibilities. (Plugge pronounced his name 'Plooje', claiming Belgian-Dutch origins; it was only when he later stood for Parliament that he agreed to the slogan 'Plugge in for Chatham' and accepted the way almost everybody else pronounced his name.)

Plugge was a pioneer of long motoring holidays on the European continent. On one such journey he stopped for coffee at the Café Colonne in the coastal village of Fécamp in Normandy. He asked the café owner what there was to see in the town and was told about the local Benedictine distillery, which was owned by the Le Grand family. Plugge was told that Fernand Le Grand had a small radio transmitter installed in his house.

Always keen to meet another radio enthusiast, Plugge went to see Fernand Le Grand at his home. During their conversation, Le Grand told Plugge that in one broadcast he mentioned the name of a local shoemaker and recommended his wares. The shoemaker's sales increased enormously. Plugge instantly saw the commercial possibilities and offered to buy time to broadcast programmes in English. Le Grand agreed and Plugge formed the International Broadcasting Company.

The fledgling organisation needed a base in London to handle advertising. So the IBC, as it became known, opened premises at 11 Hallam Street. The new headquarters were just a short distance away from the BBC's Broadcasting House, which was being built at the time.

A studio was constructed in the hayloft over the old stables in Rue George Cuvier. Old rugs draped the walls and stable matting was placed on the floor to reduce any extraneous noise. The transmitter site was moved to the Benedictine distillery's herb garden at the edge of the Normandy cliffs. The transmitter was housed in a small hut and the signal was fed to two second-hand aerial masts. This set-up provided a strong signal heard throughout London and the South of England.

Radio Fécamp started broadcasting in English on 6 September 1931. IBC's English programmes were broadcast after the French programmes had gone off the air. The first presenter, William Evelyn Kingwell, was a cashier from the National Provincial Bank's Le Havre branch, whom Plugge had met when drawing cash after leaving Le Grand. Kingwell rode over on his motorcycle on Sundays to introduce records on programmes that were broadcast from 10.30pm to 1am.

David Davies became general manager and chief announcer just as the station was renamed Radio Normandy. At this point programmes began to be transmitted every Saturday and Sunday. Many others joined during the life of Radio Normandy. Max Staniforth was an ex-army Major who was hired to be Plugge's 'man on the ground'. He moved his family to Fécamp and they would make guest appearances on air. Staniforth remained at Normandy until November 1932. He was transferred to Radio Toulouse before taking up a position at IBC's headquarters in London.

Bob Danvers-Walker began his radio career in Melbourne, Australia, in 1925, moving on briefly to ABC in Sydney in 1932 before returning to join IBC that same year. He became chief announcer at Radio Normandy and also helped establish radio stations at Paris, Madrid and Valencia.

Danvers-Walker went on to carve out a memorable career in broadcasting. He joined the BBC in 1943, and was deployed on a variety of morale-boosting wartime radio shows, including *Round and About* and *London Calling Europe*. In the 1950s Danvers-Walker was one of the regular presenters of *Housewives' Choice* and contributed to many other programmes. He also worked for Radio Luxembourg and was the announcer for the science-fiction series *Dan Dare, Pilot of the Future*. His distinctive voice was heard continuously from 1940 to 1970 as the commentator for the twice-weekly British Pathe newsreel. The upbeat, patriotic, 'stiff upper lip' style he adopted in this role soon swiftly became a standard for the medium, and is still parodied to this day whenever television or film wants to suggest 1940s and 1950s news coverage. After the arrival of commercial television to Britain in 1955, he became the announcer on Michael Miles' popular game show *Take Your Pick.*

David Newman joined the station in 1936. There was already a gentleman with the same name who worked for IBC so he became Ian Newman on air. Newman married a French girl and stayed with the station until the outbreak of the Second World War. After the war Newman joined the Foreign Office in the Diplomatic Service.

Radio Normandy was entirely financed by advertising. Philco was an early sponsor. Henleys, a car sales company, successfully launched the SS1 motorcar on the station and this demonstrated to sceptics that radio advertising really worked. At the other end of the spectrum, companies who produced affordable goods aimed at the household or medicinal markets also invested heavily. Vitamin pills, nerve tonics and mouthwashes were offered to combat such conditions such as body odour, halitosis and listlessness.

Normandy's programmes were expanded in 1932 and ran from 6pm to 3am. The power of the transmitter increased after Plugge received backing from film studio and cinema chain owner Gaumont British. A new studio was established in a house in the town. Radio Normandy by now had a large audience as far north as the English Midlands, and many big names of the day were heard. Among them was Roy Plomley, later famous for creating and presenting *Desert Island Discs* for BBC radio.

Roy Plomley took charge of an increasingly important part of the IBC Empire: outside broadcasts. The best remembered of these was *Radio Normandy Calling*. This was a touring review that first appeared in March 1938. Sponsored by MacLean's Peroxide Toothpaste, *Radio Normandy Calling* presented acts such as Alfredo and his Gypsy Band, Ward and Draper (singers), Maisie Weldon (impressionist) and Joe Young (comedian). The organisation had a fleet of trucks that traversed the country recording specially staged variety shows from cinemas and theatres. These performances were cut onto wax discs using bulky and primitive equipment aboard each van.

The recording machine was a larger version of a typical gramophone of the time with a heavy metal base. Each machine had a large electric motor at the back. A special type of cutting head replaced the typical pick-up arm of a record player. The cutting head was simply a phonograph pickup in reverse; audio was fed in and transferred to the disc on the turntable. The cutting head was moved uniformly across the disc by means of a slowly rotating feed screw mechanism.

The wax discs were big and heavy and could only be played back once. The method of playback also damaged the discs considerably. Copies could be made but this was a very costly practice. Luckily a new development was available that made the process easier and more reliable. Cecil Watts, a musician-turned-inventor, invented the lacquer-coated disc, an aluminium or glass disc layered with a cellulose-nitrate lacquer. These discs were tough enough to play back several times without wearing out. Unfortunately, the chemical composition of the lacquer has ensured that only a few of these discs survive today.

By April 1933 English programmes were being transmitted from Fécamp every day of the week. In an effort to gauge listenership, Radio Normandy created the 'International Broadcasting Club', which cost nothing to join, just the price of a stamp. Within three weeks nearly 50,000 applications had been received. Three months later more than a quarter of a million names were on the membership list.

Radio Normandy could be heard across Southern England and beyond and its output swiftly became popular. The programmes were livelier than the stodgy BBC output. On Sundays, when the BBC was concentrating on religious programmes, Radio Normandy was said to command eighty per cent of the British radio audience.

The International Broadcasting Company began leasing airtime from other radio stations in Europe and reselling it as sponsored English language programming aimed at audiences in Britain and Ireland. The IBC established a network of stations broadcasting sporadically across the continent.

Radio Toulouse started English broadcasts in October 1931 with W. Brown Constable at the helm. The first programme was *The Vocalion Concert of Newly Released Records*. Brown Constable lasted until 1933 when Tom Ronald replaced him. Ronald's tenure at Toulouse was brief, however, as broadcasting was suspended in July of that year.

When English broadcasts resumed in 1937, it was with a different company. The advertising agency W.E.D. Allen was run by three brothers who struck up a relationship with Peter Eckersley, the eccentric ex-BBC and Marconi engineer. The inaugural programme featured a speech by Winston Churchill. The main on-air personalities were Joslyn Mainprice, Allan Rose and Polly Ward. Radio Toulouse's programmes included *Feen-a-Mint Fanfare, Horlicks Picture House* and *The Empire Pools Sports Programme*. Unfortunately, the station failed to attract enough advertising and folded in May 1938.

Radio Côte d'Azur was established in April 1934. English programmes were transmitted every Sunday night and consisted mainly of gramophone records introduced by Leo Bailet. There were also regular transcribed relays from the 'Sporting Club' in Monte Carlo and the 'Coconut Grove' in Hollywood.

In 1935 the French Government announced plans to transmit from Nice and call the station Radio Côte D'Azur They demanded that the commercial station change its name. The station dutifully changed its name to Radio Méditerranée and this incarnation lasted until January 1938.

Radio Rome's first English broadcast was *A Concert of Mayfair Records* sponsored by Ardath Cigarettes. Alexander Wright was the IBC announcer. Italy was not one of IBC's successes and English shows only lasted for a couple of months until October 1934.

The Spanish stations lasted longer. The IBC began to hire time on Radio Valencia, Radio Barcelona, Radio San Sebastián and Unión Radio Madrid. Programmes from Spain started in December 1933 and lasted until March 1935. Broadcasts from these stations ended because of the rapidly deteriorating political situation that eventually led to the Spanish Civil War in 1936.

English programming on Poste Parisien started off in a small way. In 1934 English shows were broadcast for only four hours per month. However, by 1939 this had expanded to eighteen hours per week. This may have been at the instigation of Douglas Pollock. In addition to his duties as the main announcer, he was also a director of Poste Parisien. IBC was appointed as the main contractor and a lot of familiar sponsored shows were transmitted. Poste Parisien's 'state of the art' facilities included a large concert hall and two smaller studios. Station announcements were punctuated by the sound of a gong being struck. Radio Luxembourg would also adopt this practice much later.

If a station closed down another would pop up in its place. IBC programmes were also broadcast from Radio Ljubljana in Yugoslavia and Irish stations such as Dublin, Cork and Athlone. The IBC stations had twenty-one advertising sponsors in 1932; by 1935 annual revenue had climbed to £400,000 and that figure had risen to £1,700,000 in 1938.

An IBC announcer tended to lead a peripatetic lifestyle as they were transferred to different stations as and when required. Although they were paid a healthy wage, they had to work long hours for it. Homesickness was also a problem and announcers' tenures could be brief. In addition to the ones already mentioned, other staff announcers included Thorp Deveraux, Nancy Crown, Henry Cuthbertson and Hilary Wontner.

The IBC moved to bigger and better premises further up Portland Place, just about 180m from Broadcasting House. At first the BBC seemed unconcerned by this new rival, but became increasingly irritated as time passed. Firstly, there was the similar name, IBC. Even the company logos and documentation were comparable in design. When the BBC started the Empire service, Plugge countered with his own IBC Empire service on shortwave from EAQ Madrid.

The early style of 'live' sponsored programming, with a presenter simply linking records, was quickly replaced by more complex productions as radio expanded. Some of the output was imported from across the Atlantic. American favourites such as *Stella Dallas, Backstage Wife* and *Young Widow Jones* were regularly heard on Radio Normandy.

Nevertheless, nearly all sponsored shows on the continental stations were pre-recorded in Britain. As mentioned earlier, these shows were mostly recorded on gramophone records, although the fragility of these discs meant a more robust style of recording was sought. IBC's connection to Gaumont British Cinemas provided the perfect solution. Sponsored shows were recorded onto the soundtrack of film reels and played back using projection equipment.

The production of sponsored shows meant that advertising agencies had to adapt the way they operated and became programme makers as well. Agencies with only a few select clients couldn't economically establish their own production facilities so IBC quickly stepped in and offered its services for hire. The Universal Programmes Corporation, as the new division became known, produced over 2,000 programmes in the first three years of its existence.

Other advertising agencies involved in programme-making included the London Press Exchange and the J. Walter Thompson Organisation. The London Press Exchange produced programmes at the HMV recording studio at Abbey Road. Over at Bush House in the Strand, J. Walter Thompson introduced the Philips-Miller system of recording sound on metal tape. The investment was immense but continuous recording on tape swiftly showed financial returns.

Not all radio advertising took the form of sponsored programmes. For example, Ingersoll watches sponsored the time signal on Radio Normandy. There were also a certain number of 'live scripts', which were usually read by studio announcers at suitable intervals between programmes.

The elaborate scripts for Renis Face Cream utilised a storytelling format to capture the imagination of the listener. A romantic story was concocted about archaeologists unearthing a long-forgotten formula for beauty cream. The tale progressed gradually over time and listeners would hear the latest instalment each time they tuned in.

Like the BBC in its early days, the IBC encountered a great degree of suspicion and hostility from the British press. The Newspaper Proprietors Association regarded these stations as a threat to their own advertising revenues and refused to publicise them.

One exception to this ban was the *Sunday Referee*, a highly regarded sporting paper that had become a family Sunday paper. Valentine Smith ran the newspaper. He had been the circulation and publicity director of the *Daily Mail* and it had been his idea to introduce the 'Radio Yacht' to boost circulation. He realised how successful this had been and decided to involve the *Sunday Referee* in broadcasting. The newspaper ran several broadcasts on the European stations and regularly featured programme details within its pages. This was cited as one of the main reasons why it trebled its circulation figures.

The Newspaper Proprietors Association denied the *Sunday Referee* access to its transport and distribution infrastructure. Countrywide distribution was too big an undertaking for a single paper to embark upon unaided and it was eventually forced to withdraw from radio altogether.

As the IBC stations were unable to get their programme details published in the newspapers they launched their own listings magazine. *Radio Pictorial* was launched in 1934. This publication was a lot less formal than the *Radio Times*, its stuffy BBC equivalent. *Radio Pictorial* contained pictures of radio personalities and gossipy articles about the programmes. It was the 1930s equivalent of the glossy showbiz magazines of today.

Perhaps the Newspaper Proprietors Association's aggressive tactics towards the IBC and its supporters was encouraged at a higher level. The British government was openly hostile towards the continental stations. They persuaded royalty organisations to overcharge them for permission to play recorded material and encouraged the BBC to boycott any artist or presenter who had worked on a continental station. It would seem that the government was anxious to suppress any means of mass communication over which it had no control. In 1936 a committee looking at all aspects of radio stated, 'Foreign commercial broadcasting should be discouraged by every available means.'

Radio Normandy remained the flagship station of the IBC. A new studio building and upgraded transmitter site was required so building work began at Louvetot in November 1935. The new transmitter site took just over three years to complete. Broadcasts from the new studios at Caudebec began on 12 December 1938 and the new premises were officially opened on 4 June 1939. However, the new transmitter and studio fell silent on 7 September at the outbreak of the Second World War.

Plugge's IBC successfully demonstrated that state monopolies such as the BBC could be broken. Other parties became attracted to the idea of creating a new commercial radio station specifically for this purpose. One of these organisations was a French company called Radio Publicity Limited. The company chairman, Jacques Gonat, was a man of considerable influence in France. He hired time on Radio Paris and looked for advertisers in Britain.

Radio Paris was already established and very well equipped, with a power fifteen times greater than that of Radio Normandy. The first programme was broadcast on 29 November 1931 with Rex Palmer, an ex-BBC man, at the microphone. Palmer introduced *A Concert of HMV Records*.

Regrettably, Gonat's enterprise became a victim of its own success. Radio Paris attracted so many British advertisers that it incurred the wrath of French listeners. Hearing the English language so frequently on their number one station irritated them. The French government intervened and forced the English service off the air.

In 1924 Francois Aneu built a 100-watt transmitter in the Grand Duchy of Luxembourg. Because Luxembourg is centrally located in Western Europe, it was ideally placed to reach audiences in many nations, including the United Kingdom. On 11 May 1929 he brought together a group of mainly French entrepreneurs and formed the Luxembourg Society for Radio Studies as a pressure group to force the Luxembourg government to issue them a commercial broadcasting licence.

On 19 December 1929 the Luxembourg government passed a law that would allow one company to operate a commercial radio broadcasting franchise from the Grand Duchy. Ten days later the licence was awarded to Francois Aneu's society, now known as the Luxembourg Broadcasting Company. The new station would be identified on the air as Radio Luxembourg.

In May 1932 Radio Luxembourg began test transmissions directed at Britain and Ireland. This provoked a hostile reaction from the British government. The longwave band used for these tests produced a much better signal than anything previously received from outside the country. The British government accused Radio Luxembourg of 'pirating' the various wavelengths it was testing. The station had planned to commence regular broadcasts on 4 June 1933, but the complaints caused Radio Luxembourg to keep shifting its wavelength.

The English service was leased to Radio Publicity (London) Ltd and the first English transmission was fixed for 3 December 1933. This was to coincide with the last transmission from Radio Paris. So, on that occasion the two stations broadcast the same output simultaneously. Frequent announcements were made stating that all future broadcasts would emanate from Luxembourg only. Listeners were invited to retune from Paris to Luxembourg and mark the new position on the dial.

On 1 January 1934 a new international agreement, the 'European Wavelength Plan', came into effect. The Luxembourg government refused to sign the agreement, and shortly afterwards Radio Luxembourg started a regular schedule of English-language radio transmissions. The broadcasts used a new 200-kilowatt transmitter on 230 kilohertz in the longwave band. It quickly became such a significant part of the broadcasting fabric and millions favoured its laid-back style to the more formal approach of the BBC.

Stephen Williams was appointed as the first manager of the English language service. He arrived in Luxembourg with several hundred records and a couple of hampers of musical arrangements. Although only in his twenties, Williams was already a radio veteran. Prior to his appointment at Luxembourg, he had been Director of English Programmes at Radio Paris and had also worked briefly at Radio Normandy.

In the years up to the Second World War, Radio Luxembourg gained a large audience in the UK and other European countries with sponsored programming aired from noon until midnight on Sundays and at various times during the rest of the week. One of Radio Luxembourg's great successes during this period was a show aimed at children. Ovaltine, the manufacturers of a hot chocolate malt drink, first sponsored *The Ovaltineys* in 1935. The show, broadcast on Sunday evenings, was an immediate and huge success. The club associated with the show had attracted five million members by 1939; each received a membership badge and book and the chance to take part in competitions and other activities. There was a weekly comic, too, and the programme's theme song was permanently ingrained in the memories of many generations of British people.

Radio Lyons was the last of the continental stations to broadcast English programmes in December 1936. Tony Melrose was the first announcer. Radio Lyons promoted Melrose as a romantic figure and personality rather than just an announcer. In a series of magazine articles he was billed as 'the golden voice of Radio Lyons'.

Radio Lyons also heavily promoted another personality. Photographs of 'The Mystery Man of Radio Lyons' were distributed to magazines. These featured a masked figure along with details of the programme he presented. The Mystery Man's show *Film Time* was broadcast daily and sponsored by Campbell's soups.

The other output was the usual sponsored programmes and featured acts such as Carson Robinson and his Oxydol Pioneers, and Carrol Gibbons under various pseudonyms. The advertisers included Stork Margarine, Dolcis Shoes, Bile Beans and Drene Shampoo. The H. Samuel Everite Time Signal was played out between the programmes.

In the early hours of 1 September 1939 German forces invaded Poland. The world was about to be plunged into conflict and the network of European stations broadcasting to Britain faced an uncertain future. Some stations realised it would be impossible for private companies to continue and closed down immediately. Others vainly attempted to carry on until events, or the Nazis, overtook them.

There followed a period of consolidation as each European country's radio stations were gradually brought under governmental control. During the war, radio would become a powerful aide in the dissemination of information and propaganda. In addition to the physical conflict, a war of words was about to be waged for the hearts and minds of radio listeners around the world.

Chapter 6

Radio at War

With Hitler's invasion of Poland on 1 September 1939, British Prime Minister Neville Chamberlain's policy of appeasement had failed. On that day the Regional Programme was suddenly discontinued and merged with the National to become the Home Service. This attenuated service continued for the first six months of the war.

The BBC had been preparing for the outbreak of the Second World War for some months. It was feared that its transmitters might have been used as a navigational aid by enemy aircraft. The television service was also suspended for the same reason. The number of television receivers in the London reception area had risen to 23,000 by this time.

The longwave transmitter was closed and the medium wave transmitters grouped into three synchronous groups of four. All Home Service transmissions were transmitted on two frequencies (668 kilohertz and 767 kilohertz) with the network of transmitters synchronised to obstruct direction-finding capabilities.

On the morning of Sunday, 3 September 1939, Prime Minister Chamberlain announced that Britain was at war with Germany. The speech was relayed direct from the cabinet room at 10 Downing Street. King George VI's message to Britain and the Empire was broadcast that evening.

The BBC Home Service mainly concentrated on news and informational programmes in the early days of the war. There were now ten news broadcasts daily, whereas there had been five before war broke out. Nearly half the nation listened to the nine o'clock news each evening. This was because of the nightly war reports from distinguished correspondents such as Richard Dimbleby, Frank Gillard, Godfrey Talbot and Wynford Vaughan-Thomas.

In the first couple of months these reports included an eyewitness account from Edinburgh of the German air raid on the Firth of Forth,

Richard Dimbleby reporting from the British Expeditionary Force's headquarters in France and an interview with survivors of the *Athenia*, the British liner torpedoed by the Germans.

In addition there were frequent pep talks given by ministers and civil servants. Winston Churchill, First Lord of the Admiralty, made his first wartime broadcast on 1 October 1939. Countless gramophone records punctuated these talks and news bulletins. A new programme called *The Home Front* started on 30 September. This programme detailed aspects of wartime life in Britain.

Sandy MacPherson, the organist, was gainfully employed during the first fortnight of the war. MacPherson played up to twelve hours per day, also filling in with announcements and programme notes while the organisation hastily evacuated its staff from London to various locations around the British Isles. His normal signature tune was *Happy Days Are Here Again* but this would have been highly inappropriate so it was swiftly replaced by one of his own compositions.

The German radio service had also been placed on a war footing. During the Weimar Republic era, radio broadcasts had been controlled by the Postmaster General's office. In March 1933 Dr Joseph Goebbels transferred this power to the Ministry of Public Enlightenment and Propaganda with himself at the helm. Goebbels called the radio the 'eighth great power', noting its effect in helping promote the aims of the Third Reich. His broadcasts enforced the Nazi ideals of patriotism, adoration of Hitler, Aryan pride etc.

Goebbels sanctioned a directive in which millions of cheap radio sets were subsidised by the government and disseminated to German citizens. These sets were known as 'People's Receivers'. He also attempted to control what the German people listened to. On the day Britain declared war, listening to the BBC was made illegal in Germany. It was now a treasonable offence if caught doing so. In the first year of the war alone, 1,500 Germans were imprisoned for listening to London-based broadcasts.

The Nazis had been jamming BBC broadcasts experimentally for several months prior to the start of hostilities. After 3 September 1939 the jamming increased. The BBC retaliated by increasing transmitter power and adding extra frequencies. Later, leaflets were dropped over cities in occupied Europe. These contained instructions detailing how to construct a directional loop aerial that would enable listeners to hear the stations through the jamming.

The outbreak of war immediately affected staff numbers at the BBC. A lot of staff were reservists and went off to serve in the forces. In addition, each post was graded and only essential personnel were kept on; the rest were told to find alternative employment for the duration.

Although they weren't directly involved at this point, the American public were continuously being informed of the worsening situation in Europe. American journalists had been constantly reporting back home as Europe was mobilizing for war. Hans von Kaltenborn became renowned for his broadcasts from Europe, where he conversed with world leaders such as Mussolini and Adolf Hitler. His broadcasts from Spain during the civil war helped to rally American opinion against Fascism. Kaltenborn attended when the British prime minister, Neville Chamberlain, flew to Germany to meet Hitler, translating and interpreting events for the American public. The Munich crisis made him the nation's leading broadcast journalist.

Edward R. Murrow, another CBS reporter, managed to broadcast from Vienna during the annexation of Austria and went on to make gripping reports from London during the Blitz. The American correspondents reported on the distribution of gas masks in Prague, Hitler's fiery threats from Berlin and the paralysis of the French cabinet in Paris.

Radio Luxembourg's English service closed down when war broke out because the Luxembourg government wanted to avoid any accusations of bias. They had adopted a careful non-belligerent position in order to prevent a German invasion. This proved ultimately futile as the Germans eventually occupied Luxembourg in May 1940 and Radio Luxembourg's facilities were incorporated into the German broadcast network.

Most of the International Broadcasting Company stations closed down as it became obvious that war was inevitable. Radio Normandy's transmitter and studio at Caudebec was immediately mothballed and wasn't used again until the Germans utilised the facilities to broadcast propaganda to Britain. This didn't last long as the RAF bombed the transmitter out of action.

Radio Normandy's transmitter at Fécamp was used to transmit programmes from Radio International. This station's output was aimed at troops in France and the South of England during the period now known as the 'Phoney War'. It broadcast for thirteen hours a day and featured old IBC programmes with the sponsored messages removed. On Christmas Day 1939 special shows featuring stars like Charlie Kunz, Tessie O'Shea and George Formby were transmitted.

Radio International even published its own magazine, *Happy Listening*, which was distributed free to all active units of the British forces. The name of the magazine was a direct link to Radio Normandy as the station had used the slogan 'Happy Listening' in their programme listings in *Radio Pictorial*. Only two issues were published as the station had a relatively short lifespan. Radio International lasted until 3 January 1940. It's thought that pressure from the French and British governments forced the closure.

During the 'Phoney War' it became clear that members of the armed services were now mainly sat in barracks with little to do. The BBC Forces Programme was launched on Sunday, 7 January 1940 to appeal directly to these men. Its mixture of drama, comedy, popular music, features, quiz shows and variety was richer and more varied than the former National Programme.

Although intended for soldiers, the Forces Programme was also a hit with large numbers of British civilians. Programmes such as *Danger – Men at Work* and *Hi Gang!* were aimed at the forces generally. However, other shows were also developed for specific services including *Garrison Theatre* for the Army and *Ack Ack Beer Beer* for the anti-aircraft and barrage balloon stations.

The people working on the Home Front were not forgotten either. *Music While You Work* was a music programme broadcast twice daily to British factory workers on the BBC General Forces Programme. It began on 23 June 1940 following a government proposal that daily broadcasts of cheerful music piped into the factories would improve morale and increase production. The first programme featured Dudley Beaven at the theatre organ with the afternoon edition's music provided by a trio called The Organolists.

The organs were phased out gradually as it was felt their tone was incompatible with factory conditions. Small instrumental ensembles of ballroom dance orchestras, light orchestras, and brass and military bands replaced them as they were considered to be more suitable. Nevertheless, not every instrument was welcome on the show. Pizzicato violin playing was found to be inaudible due to background noise, and drummers were banned from playing 'rim shots' as these apparently sounded like gunfire over factory tannoy systems.

At first the programme had no signature tune, but light music composer Eric Coates had written a melody called *Calling All Workers* and this was adopted as the signature tune from October 1940. A third edition of the show was introduced at 10.30pm for night-shift workers. By the end of the war five million workers in over 9,000 factories were tuning in. The programme continued after the war and remained a regular broadcasting fixture until 1967.

Worker's Playtime was another programme designed to boost the nation's morale. Broadcast from factory canteens around Britain, it was one of the first touring variety shows on the BBC. Many well-known acts were given their first break on the show including Frankie Howerd, Ann Shelton, Julie Andrews, Morecambe and Wise, Ken Dodd and Bob Monkhouse. The show made its debut on 31 May 1941, and though initially scheduled to run for six weeks, it continued until 1964. Bill Gates was the producer for the entire period that the show was on air. Gates would end each programme with the words 'Good Luck All Workers'.

Popular American variety programming was imported for the first time. *The Charlie MacCarthy Show, The Bob Hope Show* and *The Jack Benny Hour* were all heard on the Forces Programme. Vera Lynn's programme *Sincerely Yours* was a big hit with the listeners and she swiftly became the 'Forces Sweetheart'. Surprisingly, the programme was less popular with the BBC Board of Governors. The minutes of one meeting dismissed the programme with the words 'Popularity noted, but deplored'.

The BBC offered broadcasting time to exiled governments to broadcast to occupied Europe. Radio-Londres was the voice of the Free French Forces, broadcasting up to five hours a day. The debut broadcast, on 18 June 1940, featured Charles de Gaulle and is often considered to be the origin of the French Resistance. He declared that France was not yet defeated and ended with the iconic statement: 'Whatever happens, the flame of the French Resistance must not be extinguished, and will not be extinguished.'

Radio Oranje was a Dutch-language radio programme on the BBC European Service which was broadcast for fifteen minutes at 9pm each day.

The first broadcast took place on 28 July 1940, when Queen Wilhelmina made a speech. In total, Wilhelmina spoke on Radio Oranje thirty-four times during the course of the war.

Radio Belgique was inaugurated on 28 September 1940 and transmitted programmes in French and Dutch. Though forbidden by the German occupiers, Radio Belgique was listened to by the majority of Belgians. Recognising the detrimental influence that these broadcasts could have, the Germans quickly started jamming Radio Belgique. They then established collaborationist radio stations such as Zender Brussel for the Dutch speakers and Radio Bruxelles for the Francophone listeners.

During the war the BBC's transmission equipment was placed under the direction of Fighter Command. The old Daventry longwave transmitter was converted to medium wave operation and joined the Home Service network. The Droitwich longwave transmitter was also converted to medium wave operation and, together with the other former National Programme transmitters, was synchronised on 1149 kilohertz and broadcast the European Service at night.

At a later date the Start Point transmitter in South East Devon was converted for use on 1149 kilohertz and, together with Droitwich, these two transmitters broadcast the European Service, leaving other transmitters available for a new third service to be added later.

The synchronisation of the Home Service transmitters onto just two frequencies caused numerous interference problems for domestic listeners, with one Home Service transmitter interfering with another on the same frequency. To surmount this problem the BBC installed a group of low-power relay stations around the country using 1474 kilohertz called 'Matrix H', which was later extended. This network of low-power relays filled in the coverage gaps from the main transmitters. All of the Group H stations were manned twenty-four hours per day so the broadcasts could be terminated at a moment's notice if air raid conditions warranted it.

In September 1941 the Daventry and Droitwich transmitters, together with a new transmitter installed at Brookman's Park, were established as a longwave network to broadcast the European Service on 200 kilohertz. As the war progressed, further transmitters were opened to boost reception of all the BBC services.

Extra European Programme transmitters were built at Crowborough in Sussex and a high-power transmitter at Ottringham in the East Yorkshire Riding was added in February 1943. The Ottringham transmitting station was a massive installation consisting of six 150-m-high masts and 800-kilowatt transmitters.

The BBC deployed a lot of staff to other areas of the United Kingdom to avoid frequent bombing raids, though this was not a foolproof plan as several BBC installations also came under attack. On 19 November 1940 the Adderley Park transmitter in Birmingham was totally destroyed and several staff members sheltering nearby were killed.

The expansion of the Overseas Services, the relocation to areas outside London and the transfer of personnel to HM Forces meant that the BBC Engineering Department was severely understaffed. They were forced to train school-leavers and female members of staff to fill dozens of vacant positions.

The BBC staff members that remained in London were extremely vulnerable to attack as the German *Blitzkrieg* intensified. In October 1940 a delayed-action bomb killed seven people, injured many others and blew out part of the west side of Broadcasting House. Listeners to the nine o'clock news heard the announcer pause, and then continue.

In December 1940 a landmine caused so much damage to Broadcasting House that the European service had to be moved temporarily to Maida Vale. The following April all premises on the eastern half of the Broadcasting House site were totally destroyed by high-explosive bombs. A month later Queen's Hall was demolished during a raid and the Maida Vale studios suffered a direct hit by a high-explosive bomb.

Eventually a new home for the BBC's European service was found. This was Bush House, where there were already some basement studios and offices. This had previously been the production home of advertising agency J. Walter Thomson during the 1930s.

Bush House stands between Australia House and India House on the semi-circular island of buildings called Aldwych, in Westminster, London. This quintessentially British building was originally constructed by an Anglo-American trading organisation headed by Irving T. Bush, after whom the building is named. At the time it was built it was declared the 'most expensive building in the world', having cost around £2 million. Designed by Harvey Corbett, Bush House was built in 1923 and opened in 1925. Further wings were added between 1928 and 1935.

All of the BBC's foreign language services gradually invaded Bush House, permeating each wing in turn. Bush House also suffered bomb damage from a flying bomb in June 1944. The statue representing America lost its left arm and was restored in 1977. It was the home of the BBC World Service for seven decades but all staff transferred to Broadcasting House when the BBC's lease expired in 2012.

Although the BBC retained a degree of independence, the organisation had to adhere to Ministry of Information 'guidance'. This censorship took two forms: one covered the morale of the nation and the other protected defence forces' security. Any script had to have the two official stamps before it was broadcast. The locations of troops, cabinet members or the Royal Family were never given. Weather forecasts were banned, as this would reveal suitable conditions for bombing.

The seeming failure of the British government, including the military failure in Norway in 1940, meant that criticism of Neville Chamberlain became more and more forceful. Chamberlain stood down and Winston Churchill became prime minister on 10 May 1940. He swiftly emerged to be the most dominant figure in British politics during the war. For many people, his stand summarised why the war was being fought.

His radio speeches are often credited with helping the Allied forces to victory. However, rumours later emerged that a stand-in had delivered many of his speeches as he was too busy elsewhere. This had always been denied by official sources, but a BBC 78rpm record was later unearthed to back up the story. Shortly before his death in 1980, the actor Norman Shelley revealed that he had been the substitute for Churchill. To confuse matters further, Churchill agreed to record some of his most memorable lines for the BBC after the war, and it's these speeches that are retained in the official BBC archives.

The BBC Home Service had been implemented in a hurry and many of the pre-war favourite programmes had been dropped. For a while the Home Service broadcast a mixture of programmes but eventually the news programmes remained on the Home Service while the music and light entertainment shows transferred to a new service.

It's That Man Again

The most popular British comedy programme during the war years was *ITMA* starring Tommy Handley, a well-known Liverpudlian comedian. Handley was a veteran radio performer whose debut broadcast was a relay of the Royal Command Performance of December 1923. The show's name was an acronym of a topical catchphrase of the time. In the run up to the war, whenever Hitler made some new territorial claim, the newspaper headlines would proclaim 'It's That Man Again'.

The programme began on 12 July 1939 with a series of four fortnightly shows. The first episodes were based on board a pirate commercial radio ship. Yet these initial shows received a lukewarm response and a new format was quickly sought. In the early days of the war, new government ministries sprang up overnight. So for the second series Tommy Handley became Minister of Aggravation and Mysteries at the Office of Twerps.

ITMA returned on 19 September 1939 for a weekly series of twenty-one episodes. These were transmitted from Bristol, where the BBC Variety Department had moved, hoping to avoid the heavy bombing raids directed at London.

Handley was the only survivor from the original series and a brand new supporting cast was assembled. These included Vera Lennox as Handley's secretary, Dotty, and Maurice Denham as Mrs Tickle, the office cleaner. Sam Costa and Jack Train provided other characters including Funf, the elusive German spy. The catchphrase 'this is Funf speaking' found its way into many private telephone calls over the next few years. This series became very popular and the show was swiftly recommissioned.

Before the third series started the BBC Variety Department had moved yet again to Bangor in North Wales. The war situation was deteriorating and it was thought that spoofs of government departments would not be acceptable. The show began a six-week summer season on 20 June 1941 and was renamed *It's That Sand Again.* It was set in a sleepy seaside resort called Foaming at the Mouth. In the new series Tommy Handley became the town's mayor.

There was another change of cast with only Jack Train returning. In came Sydney Keith, Horace Percival, Dorothy Summers and Fred Yule. Several famous characters were launched during this short run. These included Lefty and Sam the gangsters and Deepend Dan the Diver. Other popular characters included Claude and Cecil, the over polite handymen, and Ali Oop, a shady Middle Eastern hawker.

The show reverted to its original name for the fourth series and the popular seaside setting was retained. In this series the team were joined by Dino Galvani as Tommy Handley's Italian secretary, Signor So-So, and Clarence Wright as a hapless commercial traveller who never made a sale. Dorothy Summers introduced the famous office cleaner, Mrs Mopp. Her catchphrase 'Can I do yer now sir?' became immensely popular.

The next three series of *ITMA* continued in the same vein apart from a few minor cast changes. During this period a film version of the show was produced and was received warmly. The cast were also invited to perform a special show at Windsor Castle to celebrate Princess Elizabeth's sixteenth birthday. One of the newer characters to appear during this period was Colonel Humphrey Chinstrap. The colonel would turn almost any innocent remark into the offer of a drink with his catchphrase 'I don't mind if I do'.

ITMA returned for its seventh series in October 1943 with another revamp. Handley was now Squire of Much Fiddling. Over the next year several special editions were recorded at various military bases around the country.

The show was immensely popular, with nearly forty per cent of the population listening at times. *ITMA* continued after the war but the series ended abruptly after the scheduled broadcast on 6 January 1949. Tommy Handley passed away three days later and the BBC wisely decided to let the show die with him.

The war had forced a change to the BBC's pre-war policy of broadcasting only serious or religious programmes on the Sabbath. Looking at the programme listings for May 1941 showed that over three hours of light entertainment shows were broadcast on Sunday evenings. There would be no reverting back to Reith's Sunday policy after the war.

The BBC commenced its 'V for Victory' campaign at midnight on 20 July 1941. The 'V for Victory' broadcasts started with a message from Prime Minister Winston Churchill aimed at the European countries under occupation by the Nazis: 'The V sign is the symbol of the unconquerable will of the people of the occupied territories and a portent of the fate awaiting the Nazi tyranny.' From that point on the BBC's broadcasts used a call sign that utilised the opening bars of Beethoven's Fifth Symphony, which has the same rhythm as the Morse code for the letter V (dot-dot-dot-dash).

On 19 February 1941 the German *Luftwaffe* bombed Swansea. Up to seventy enemy aircraft dropped some 35,000 incendiaries and 800 high-explosive bombs over three nights. The raging fires could be seen from the other side of the Bristol Channel in Devon. On the last night of the Blitz, the BBC's premises on Alexandra Road was struck and completely destroyed. At this point only the West Wales Singers' weekly service, *Y Gwasanaeth Gosper*, was being transmitted from the city. Luckily the staff had just finished their weekly broadcast when the *Luftwaffe*'s final onslaught began and escaped unharmed.

When the American servicemen arrived en masse in 1943 and 1944, in preparation for Operation Overlord, they found even the livelier Forces Programme shows to be staid and slow compared with the existing output of the American networks. Therefore an alternative was sought.

On 4 July 1943 the American Forces Network (AFN) was introduced on 871 kilohertz. Using BBC emergency facilities at 11 Carlos Place in London, the signal was fed by landline to five regional transmitters. The initial broadcasts included less than five hours of recorded shows, a BBC newscast and a sportscast.

AFN provided a morale-boosting service of record programmes that was popular with the American forces. Yet AFN was equally popular with British audiences, who could hear records and music not usually heard on the BBC. During the next eleven months the daily broadcasts expanded to nineteen hours. Fifty additional transmitters were installed, including six in Northern Ireland. Six additional staff joined the original complement of seven broadcasters and technicians.

It was the popularity of AFN and the increasing numbers of American forces based in Britain that persuaded the BBC to 'fine tune' their Forces Programme. Finally, in response to a direct appeal from General Eisenhower, the BBC abolished the Forces Programme and established the General Forces Programme on 27 February 1944.

The new service's output was lighter and less formal than that of its predecessor, with material that would be popular with American and Canadian troops. The station assumed a more North American style and played more American material. American shows like *The Bob Hope Show, Amos 'n' Andy* and *Command Performance* were broadcast.

They soon found that this new approach was a big hit with everyone, not just the troops, and certainly helped the listeners endure this troubled period. Regulars such as Canadian announcer Charmian Sansom,

Joan Dallas and the Robert Farnon Canadian Army Band became firm favourites.

AFN London moved from its original BBC studios to 80 Portland Place in May 1944. As D-Day approached, AFN combined with the BBC and the Canadian Broadcasting Corporation to form the Allied Expeditionary Forces Programme. The AEF programme made its debut on 1052 kilohertz at 6am on 6 June 1944. The signature tune *Oranges and Lemons* was followed by the opening announcement from Margaret Hubble. The first show was a music programme called *Rise and Shine*. US Sergeant Dick Dudley and AC2 Ronnie Waldman co-hosted the show.

AFN and BBC personnel accompanied the invasion force when Allied troops landed in France. They sent news reports back to studio locations in London. In addition AFN established a network of mobile stations to broadcast music and news to the troops in the field.

AFN mobile stations were set up at Paris, Nice, Marseilles, Rheims, Le Havre, Cannes and Biarritz as Allied troops advanced through Europe. AFN's administrative headquarters remained in London but its operational base moved to Paris. When Germany surrendered on 8 May 1945, the network had grown to some 700 people and sixty-three stations scattered throughout Central Europe.

There was another radio service backed by the Americans. The American Broadcasting Station in Europe (ABSIE) began on 30 April 1944, five weeks before D-Day. The Office of War Information and the Supreme Headquarters Allied Expeditionary Force's Psychological Warfare Division established it. ABSIE broadcast from underground studios in Wardour Street in Soho. The station's brief was to provide 'the truth of this war to our friends in Europe – and to our enemies.' Colonel William Paley, peacetime head of CBS, was drafted in to run the operation. He negotiated with the BBC for equipment and pledged that ABSIE would disband ninety days after V-E Day.

The station provided news and information in seven languages. There was a news bulletin presented at dictation speed for use by the underground press in occupied countries. It also featured talks by exiled leaders as King Haakon of Norway and Jan Masaryk of Czechoslovakia.

Not surprisingly, ABSIE's most popular programmes were music shows. According to captured Germans, the favourite Allied programme heard in Germany was *Music for the Wehrmacht*. This show featured top American acts like Dinah Shore, Glen Miller and Bing Crosby introducing their songs by reading from phonetic German scripts. ABSIE only lasted for fourteen months. Colonel Paley fulfilled his promise to the BBC and closed the station shortly after V-E Day.

The BBC's German Service had continued to broadcast throughout the war and by 1942 was broadcasting around six hours per day. They had employed a policy of restraint in the early years of the war and took great pains to maintain a level of neutrality. This approach began to pay dividends and audience levels steadily rose.

The German Service adopted a relaxed and calm tone for its broadcasts. This broadcasting style was more akin to pre-war Radio Luxembourg than the BBC Home Service. The announcer's relaxed manner contrasted sharply with the frenetic style of the Nazi broadcasts.

The most-listened-to programme on the BBC's German service was transmitted every afternoon. Spike Hughes introduced *Aus der Freien Welt,* which translates as *From the Free World.* He played hot jazz and swing records that were banned in Germany. The records were interspersed by news and short talks.

The German Service also featured monologues and sketches. One of the more popular series featured a character called Adolf Hirnschal, a German private, writing letters home to his wife. An Austrian refugee actor and writer called Robert Ehrenzweig, who became known as Robert Lucas in Britain, wrote the scripts for these monologues. Other popular radio characters on the German Service included *Blockleader Brownmiller* and *Mrs Wernicke*. Bruno Adler, another German writer in exile, created these radio spoofs.

Radio as propaganda

Radio can be a highly effective propaganda tool and both sides utilised it extensively during the war. There are essentially three different types of propaganda: white, grey and black. White is the most common type of propaganda where the real source is declared. Usually more accurate information is given, if also slanted or distorted. With grey propaganda, the source is never identified. Black propaganda is false information and material that claims to be from a source on one side of a conflict, but is really from the opposing side. It is typically used to vilify, humiliate or misrepresent the enemy. These black propaganda radio stations often pretended to broadcast illegally from within the countries they targeted.

The British government launched several black propaganda stations during the Second World War. These operations were the responsibility of the Political Warfare Executive based near Woburn Abbey. This unit produced black propaganda radio stations in most of the languages of occupied Europe.

Amongst the various language services, two were aimed at Italy. These were Radio Italy and Radio Liberty, which revealed curious intimate details of Mussolini's private life. The Italian stations were suspended when it quickly became apparent that the Italian farmers and workers didn't have shortwave radios and therefore couldn't receive the broadcasts.

Not surprisingly, most of these clandestine stations broadcast in the German language. The mastermind behind this unit was Sefton Delmer, who had been the Berlin correspondent of the *Daily Express* prior to the outbreak of the war. He was able to speak German perfectly and had specialist knowledge of the Nazis. He had met most of the top party members, including Hitler, during his time in Berlin.

Delmer's first effort was a right-wing shortwave station called Gustav Siegfried Eins (George Sugar One). To front this station they invented a character called *Der Chef* (The Chief). This individual professed to be a Nazi extremist who accused Hitler and his henchmen of going soft. The station focused on alleged corruption and included salacious stories regarding the sexual proclivities of Hitler Youth Leaders. There were also frequent stories concerning the outrageous sexual behaviour of Nazi party bosses with the wives of absent *Wehrmacht* troops.

The Political Warfare Executive created a grey propaganda station for the Navy. This was the station *Atlantiksender*, broadcasting non-stop music and news for the U-boat crews. Although most of the U-boat crews realised that *Atlantiksender* had Allied origins they became genuinely troubled at just how much the enemy knew. With a growing first-rate intelligence department, daily intelligence reports from the Navy and the Combined Services, they gained valuable details to add to their stories.

These stations were designed to demoralise and spread dissension amongst the German troops. Information was gathered from various sources such as secret service reports, censorship intercepts, newspaper items and RAF reconnaissance reports. They also used information and names of actual people gleaned from captured German sailors, soldiers, and airmen. Actual news was interspersed with fake items. One false report revealed that tests had shown that a large quantity of blood used in German army field hospitals was contaminated with syphilis.

On 8 November 1942 the British installed a high-power medium wave broadcasting transmitter near Crowborough in the Ashdown Forest in Sussex. It was nicknamed Aspidistra after the Gracie Fields song *The Biggest Aspidistra in the World*. Aspidistra could switch frequencies and wave bands almost instantly and was used extensively for propaganda broadcasts.

Despite its strategic importance, the Crowborough transmitter never suffered a direct attack by German aircraft. On one occasion an aircraft offloaded its cargo of incendiary devices a mile from the site. However, this was thought to have been a standard raid and Aspidistra was not the main target. There was another incident when a V1 rocket skimmed over the hill at Crowborough, which was the highest point before London. The doodlebug flew under the guy-wires of the mast and continued on its way to the capital.

Delmer's next project was a twenty-four-hour broadcasting station on medium wave. *Soldatensender Calais* (Calais Armed Forces Radio) was introduced just a few weeks before the invasion of France. *Soldatensender* brought the first news of the D-Day landings to the world. The communication breakdown between German units at this point was so severe that many commanders tuned in to the station for situation reports. They would use these reports to correct the constantly changing battle plans on their staff maps. This information was correct 99 times out of 100, but on the hundredth it would drive the Germans into a trap set by the allies.

The attempted assassination of Hitler was exactly what the Woburn Abbey team had been waiting for. They maximised the opportunity by implicating as many as possible in the plot. Emboldened by this development, the *Soldatensender* demanded that an end be put to the war to save Germany.

As the Allied forces advanced across Europe, Radio Luxembourg and the BBC were telling the German civilians to 'stay put'. Winston Churchill was outraged by this and instructed the black propagandists to try to unsettle the German people and cause as much disruption as they possibly could.

Soldatensender broadcast a bogus report saying that seven bomb-free zones had been established in central and south Germany. It stated that these areas would be supervised by neutral Red Cross representatives and refugees would be safe from further enemy air attacks. Sending the civilians on to the roads of Germany would disrupt supplies and block the German army's retreat.

Also, with first-class intelligence information and using advanced knowledge of Allied air raids, Delmer's team were able to predict which station would go off the air, and when. As the German station went silent they would take over the German network. Specially trained announcers would make bogus announcements identical in rhythm and tone to the genuine station.

The Nazis also realised the potential of propaganda broadcasts. Just a few short months after the outbreak of the war, the Germans were transmitting close to eleven hours of programming a day. *Germany Calling* was broadcast to audiences in Great Britain on the medium wave station *Reichssender Hamburg* and by shortwave to the United States of America. The Nazi transmissions gave prominence to reports on the sinking of ships and the shooting down of aircraft. German radio frequently rang a mock lutine bell to announce the sinking of a British vessel.

These broadcasts stemmed mainly from Berlin but Radio Luxembourg's transmitters were also used to relay them. In an effort to prevent their transmissions from being bombed out, the Nazis had perfected an ingenious network of mobile transmitters. An extensive system of landlines was located along all the major routes in Germany; every few kilometres there was a concrete box where engineers could plug in their equipment. Mobile broadcasting units stored in the back of trucks were moved around like chess pieces to these connection points. They would stay put for a few hours and move on before the Allied bombers could pinpoint their location.

Although the broadcasts were widely known to be Nazi propaganda, Allied troops and civilians frequently listened to them. Despite the inaccuracies and exaggerations, they frequently offered the only details available from behind enemy lines concerning the fate of friends and relatives who didn't return from bombing raids over Germany.

Doctor Joseph Goebbels assembled a team of broadcasters to promulgate his disinformation to all the combat zones. Mildred Gillars, an American expatriate, was one of those propagandists.

She was working for German State Radio when hostilities between the United States and Germany began. Gillars elected to stay in Germany with her fiancée and was enlisted to present several programmes targeted at the American troops.

One of her programmes, *The Home Sweet Home Hour*, was aimed at making the fighting men feel homesick. On another programme, *Midge at the Mike*, she played American songs interspersed with defeatist statements, anti-Semitic rhetoric and verbal attacks on Franklin D. Roosevelt. She soon gained several nicknames amongst her audience, including 'Olga' or 'the Berlin Babe'. The one that stuck was 'Axis Sally'. This name possibly transpired when she was asked to describe herself on air, Gillars had said she was 'the Irish type … a real Sally'.

After the war Gillars was arrested and charged with treason. In 1949 she was sentenced to ten to thirty years in prison. She became eligible for parole in 1959, but wasn't released until 1961.

For British audiences, the most prominent Nazi propaganda broadcaster was William Joyce, who was Irish by blood, American by birth and carried a British passport. His focus was on eroding the morale of the British people, who nicknamed him Lord Haw-Haw. Joyce had been interested in politics from an early age. Both he and his father, rather unusually for Irish Catholics at the time, were both Unionists and openly supported British rule.

In 1924, after attending a Conservative Party meeting at the Lambeth Baths in Battersea, Joyce was accosted by a gang and subjected to a vicious attack. He received a deep razor cut that sliced across his right cheek from behind the earlobe all the way to the corner of his mouth. After spending two weeks in hospital he was left with a disfiguring facial scar. Joyce was convinced that his attackers were 'Jewish communists' and the incident became a major influence on the rest of his life.

He joined the British Union of Fascists in 1932 and fled to Germany just before the war broke out in 1939 (Joyce had been tipped off that the British authorities intended to detain him as a Nazi sympathiser). On arrival in Germany, he immediately joined Joseph Goebbels' Propaganda Ministry and broadcast weekly from 1939 to 1945.

Many of Joyce's stories were designed to unnerve the British public. On one occasion, Joyce asked the British public to question the Admiralty over the aircraft carrier *Ark Royal*. In fact, nothing had happened to the *Ark Royal* but the seeds of doubt had been sown. It is thought that on average six million people listened to Joyce's broadcasts. Many found them so absurd that they were seen as a way of relieving the tedium of life in Britain during the war. Joyce became a figure of fun: comedians lampooned his idiosyncratic accent, and there was even a song written about him, *Lord Haw-Haw, the humbug of Hamburg*.

The public may not have taken Joyce's broadcasts seriously, but that was not the case in certain areas of officialdom. A report released in January 1940 showed that, during the previous month, thirty per cent of the adult population had listened to the broadcasts from Hamburg. The BBC came under pressure from senior politicians and military top brass to reply to Joyce's claims, but the BBC's official reply was always the same: they thought that 'a permanent, regular refutation of German lies was not possible or desirable'.

On the night of 30 April 1945 a drunken Joyce made his last broadcast from Hamburg as British troops entered the city. With his adopted world crashing down around him, but still committed to the Nazi cause, Joyce rambled on through his farewell speech. In Berlin, Hitler was simultaneously saying goodbye to his entourage in anticipation of ending his life a few hours later.

Captured by the British, Joyce stood trial for treason. The court rejected his claim of American citizenship because he held a British passport. He was found guilty and hanged at Wandsworth Prison on 3 January 1946.

The myth of 'Tokyo Rose'

Lord Haw Haw's Japanese equivalent was Tokyo Rose whose mission was to demoralise the American troops in the Pacific. She was the evil seductress who urged GIs to abandon their war against the Imperial Japanese war machine.

As a matter of fact 'Tokyo Rose' didn't exist. There wasn't actually a female broadcaster using that particular soubriquet on the air, though there were at least eight women, and quite possibly more, who broadcast from Radio Tokyo, including 'Nanking Nancy', 'Radio Rose' and 'Madam Tojo'. The occupying forces under MacArthur identified five women as possibly being 'Tokyo Rose' within days of entering Tokyo.

The woman who was ultimately branded as 'Tokyo Rose' by the American press was Iva Toguri, an American born to Japanese immigrant parents. Her family didn't speak Japanese in the home, nor did they adhere to Japanese customs or eat Japanese food. She had been sent to Japan to care for an ailing relative just before the outbreak of the Pacific War. For a number of reasons she had left without proper permits and found it impossible to return to America when war was declared. She was trapped in Japan, unable to speak or read the language.

Despite pressure by the Japanese authorities, Iva Toguri refused to relinquish her American citizenship. She ended up working as a typist at Radio Tokyo. As a native English speaker it was her job to edit the scripts being prepared for broadcast. Eventually Toguri was pressurised into become one of the announcers. In wartime Japan nobody disobeyed direct orders from the army.

Iva Toguri's show was called *Zero Hour*. She had a light presentation style that was hugely popular with the GIs and her show quickly became the most popular radio show in the Pacific theatre. She would introduce herself as 'your favourite enemy, Orphan Ann.' Her comical delivery was deliberate, and removed any perceived threat from her script. Far from being a covert disseminator of misinformation, Iva openly warned her listeners that her programme contained 'dangerous and wicked propaganda, so beware!' The results were more amusing than disheartening. Instead of lowering morale, she boosted it.

After the war Iva Toguri was arrested and tried for treason. She was sentenced to ten years in prison and stripped of her prized US citizenship. Gradually, as the anti-Japanese attitude of the post-war years began to ebb, it became clear that Iva Toguri had been treated harshly. She eventually received a Presidential pardon in 1977, and her cherished citizenship was restored.

These propaganda stations gradually disappeared as the war neared its conclusion. In Europe, as the German forces retreated and countries were repatriated, the need for them evaporated. The most powerful, *Soldatensender Calais*, closed down finally on 30 April 1945, with no formal closure announcement made.

Sefton Delmer advised his team at Woburn Abbey to remain tight-lipped about their work with the Political Warfare Executive. He didn't want to offer the Germans the opportunity to claim they had been beaten by underhanded means and not militarily, though it's debatable whether these propaganda broadcasts had any real impact on the outcome of the war.

On 1 September 1944, five years to the day since the English service of Radio Luxembourg closed down, the Germans abandoned the facilities at the Villa Louvigny. They blew up the main control room in the basement of the building but left the transmission site intact. They only destroyed the transmission valves but had left the other equipment alone. A stock of transmission valves was later discovered at the Post Office warehouse at Diekirch.

Several days later the Allies regained control of the studio and transmitter facilities. The area around the transmitter site had been mined and a tank was destroyed as it approached. Eventually the transmission facilities were cleared of all booby traps and Radio Luxembourg was handed over to the Allies.

The SHAEF (Supreme Headquarters Allied Expeditionary Force) took over the running of the station. Programmes were strategic in nature and aimed at the German units still active in the combat zone. Programmes in the local Luxembourg dialect were introduced in October 1944. Eventually the Luxembourg transmitter relayed programmes from the BBC and ABSIE.

The Luxembourg facilities were also used to broadcast a black propaganda station called 'Radio Twelve Twelve' run by a team of Twelfth Army personnel. Four fifteen-minute programmes were broadcast daily. The station purported to be a German underground station and maintained an anti-Nazi stance. Its aim was to spread disinformation and unsettle the German population. It falsely reported that tanks were in the vicinity of Ludwigshafen and Nuremberg and this caused widespread panic. Radio Twelve Twelve was in existence for only 127 days but it had succeeded in its aims.

As the war in Europe began to go the Allies' way there were increasing signs that life in Britain was improving. Programmes were interrupted frequently during the autumn of 1944 to bring news of the liberation of Paris, Brussels, Luxembourg, Athens and Belgrade. The chimes of Big Ben were reinstated. News bulletins carried details of the end of fireguard and Civil Defence duties in certain areas. On 3 December the BBC news announced that the Home Guard was to be disbanded.

At 3pm on 8 May 1945 a victorious Winston Churchill officially announced the end of the war with Germany. Speaking from the Cabinet room at Downing Street, he reminded the nation that Japan had still to be defeated but that the people of Great Britain 'May allow ourselves a brief period of rejoicing. Advance Britannia. Long live the cause of freedom! God save the King!' That evening at 9pm King George VI broadcast to the nation from Buckingham Palace. This was the last official event on V-E Day.

The cessation of hostilities in Europe didn't mean an immediate end to the BBC's General Forces Programme. It moved to shortwave and continued to transmit to areas that were still engaged in active combat. Gradually, Britain began to disengage from each fighting area as civilian rule was restored. The General Forces Programme was slowly replaced by the BBC General Overseas Service until complete closure on 31 December 1946.

Chapter 7

The End of the Golden Age

Following the cessation of hostilities the BBC immediately restructured its services. The provincial regional programmes resumed but the Regional Programme itself didn't. The BBC reintroduced the six pre-war regional services, on the same transmitters and frequencies as before, retaining the wartime name BBC Home Service. So, for example, the Scottish regional station on medium wave was now called the Scottish Home Service. This was the first part of the BBC's post-war restructuring of radio into three networks.

The second phase was implemented on 29 July 1945. The BBC Light Programme commenced broadcasting on the longwave frequencies that had previously carried the General Forces Programme. The Light Programme also had the use of nine medium wave transmitters, giving full national coverage. Launch day programmes included the ubiquitous Sandy McPherson at the organ and an afternoon performance by the Torquay Municipal Orchestra.

The longwave signal was transmitted from Droitwich in the English Midlands and gave fairly good coverage to most of the UK. Some medium wave frequencies were added later, using low-power transmitters to fill in local blank spots.

The new service was generally well received, though some listeners regretted that the American acts they had listened to on the Forces Programme had been dropped. Several long-running programmes were gradually introduced including *Woman's Hour, Dick Barton Special Agent* and *Housewife's Choice.*

Responding to the public's growing interest in the arts, the BBC introduced the Third Programme on 29 September 1946. Broadcast initially for six hours each evening, the Corporation defined the new service as 'being for the educated rather than an educational programme.' It gradually became one of the leading cultural and intellectual forces in Britain, playing a crucial role in disseminating the arts.

At first, the Third Programme used 583 kilohertz from Droitwich. However, as this was not an internationally cleared frequency, coverage was limited. Therefore, twenty-two of the old Matrix H transmitters were brought back into service to supplement coverage outside the Midlands.

The Third Programme was given a unique freedom from form and routine. No news bulletins or fixed periods were allowed to interfere with the output. Plays, operas and concerts were given in their entirety. The new station became a major patron of the arts with music filling a third of its output and providing a wide range of serious classical music and live concerts. Jazz music was also heavily featured. The new network also played a crucial role in the development of contemporary composers such as Benjamin Britten by commissioning new works.

Speech formed a large proportion of the Third Programme's output. Its drama productions included plays by Samuel Beckett, Harold Pinter and Joe Orton. It also commissioned new plays by leading writers, probably the most celebrated of which was *Under Milk Wood* by Dylan Thomas.

The Third Programme was the single largest source of copyright payments to poets for many years. Young writers such as Philip Larkin, David Jones and Laura Riding received early exposure on the network. The radio talk, a staple of the early days of the BBC, also made a comeback. Nationally known intellectuals such as Isiah Berlin, Fred Hoyle and Bertrand Russell were frequently heard talking on philosophy or cosmology.

The esoteric character of the service turned out to be harmful rather than advantageous to the BBC. Although normally exalted for its noble objective of raising cultural standards, the Third Programme's high production costs and small audience figures made the service an easy target for critics and politicians.

In fact, the Third Programme's existence went against the principles of its founding father. John Reith had been opposed to fragmenting audiences by splitting programming genres across different networks. From the outset, though, it had influential supporters: the Education Secretary in the Attlee government, Ellen Wilkinson, spoke rather optimistically of creating a 'third programme nation'.

Despite its influential supporters, and following an internal reorganisation, the cultural programmes were reduced in hours in 1957. The Third programme had to share its frequency with other programming strands. This situation continued until the launch of the

BBC Music Programme on 22 March 1965, which broadcast classical music between 7am and 6.30pm daily.

While the Third Programme attracted a niche audience the other BBC services delivered listeners en masse. Programmes like *Much-Binding-in-the-Marsh, Take it from Here* and *Ray's a Laugh* became hugely popular with the British public.

Much-Binding-in-the-Marsh starred Kenneth Horne and Richard Murdoch as senior staff in a fictional RAF station. Over the years the station became a country club and finally a newspaper. The show had a chequered history and was broadcast on BBC radio from 1944 to 1950. It transferred to Radio Luxembourg between 1950 and 1951. Then it returned to the BBC until it ended its run in 1954. One of the most fondly remembered parts of the show was the closing theme tune, which featured topical lyrics sung by members of the cast.

Take It from Here ran on the BBC between 1948 and 1960. It starred Jimmy Edwards, Dick Bentley and Joy Nichols. Nichols left the show in 1953 and was replaced by June Whitfield and Alma Cogan. The show's writers were Frank Muir and Denis Norden, who are credited with reinventing British post-war radio comedy. It was one of the first shows with a significant segment parodying film and book styles but the show is probably best known for a series of sketches featuring an uncouth, dysfunctional family called 'The Glums'.

The Glum family were the antithesis of cosy middle-class families portrayed in the media at the time. Pa Glum (Edwards) was a cantankerous old soak. His son, Ron (Bentley), was a feckless layabout, despite the relentless efforts of his ambitious fiancée, Eth (Whitfield). Mrs Glum, the family matriarch, was often heard incoherently in the distance. Singer Alma Cogan usually provided Ma Glum's off-stage noises.

Ray's a Laugh, starring Ted Ray, was a more traditional show. It was essentially a domestic comedy with musical items. Kitty Bluett played Ray's wife and Fred Yule played his brother-in-law. The supporting cast included a fine array of comedy stalwarts such as Patricia Hayes, Kenneth Connor and Pat Coombs. Another early cast member was a talented newcomer called Peter Sellers, who would go on to more anarchic humour later in his career. *Ray's a Laugh* ran from 1949 until January 1961.

The year 1947 was an important one for the BBC but it didn't start too well for them. A hard winter and a fuel crisis forced the Corporation to take drastic action. They suspended the Third Programme and amalgamated the Light Programme and Home Service for a short period during February and March. Full daytime programming was not reinstated until April.

On 14 November the BBC celebrated its Silver Jubilee by publishing a pamphlet called *Twenty-Five Years of British Broadcasting*. They also transmitted a number of special programmes. On 20 November the BBC broadcast the wedding of Her Royal Highness Princess Elizabeth to Lieutenant Philip Mountbatten at Westminster Abbey.

The International Broadcasting Company (IBC) stations didn't return after the war. Changes in French media law meant that private stations couldn't operate with the same freedom as they had done before the war. In 1947 discussions were held regarding the possible revival of Radio Normandy's English Service. However, these talks were never concluded satisfactorily and the service never returned. Instead, the IBC headquarters in Portland Place became one of the best independent recording studios in Britain and during the 1960s and 1970s, they were used by some of the biggest recording artists in the world.

Radio Luxembourg remained in the hands of Allied forces for some time after the war. Special programmes were broadcast for the American occupying forces. It was also strongly rumoured that Winston Churchill wanted to use the transmitter facilities as a propaganda weapon against the Soviet Union. This plan never came to fruition and Radio Luxembourg was handed back to its owners during the summer of 1946.

When Luxembourg returned it was business as usual and Stephen Williams resumed his role as manager of the English service. Transmissions had restarted on the same pre-war longwave frequency. Unfortunately, Radio Luxembourg found it difficult to attract English advertisers and the broadcasts were gradually reduced and replaced by other languages.

On 2 July 1951 the English programmes were transferred to a less powerful medium wave frequency of 208m. These transmissions could only be received satisfactorily in Britain during the hours of darkness, when the signal was able to bounce off the ionosphere and reach the United Kingdom.

Geoffrey Everett was in charge by this time and the station adopted the slogan '208 – Your station of the stars', which referred to the performers heard on the station. Despite the relegation to medium wave, this became a golden period for Radio Luxembourg. Popular programmes at the time included *Dr Kildare, Dan Dare – Pilot of the Future* and the *Top Twenty Programme*. The latter was a show featuring the best-selling records of the week and was the first time such a programme had been attempted in Europe. *The Top Twenty Programme* became Radio Luxembourg's most popular show and received around 1,500 letters a week. Sponsored programmes made up the majority of airtime. Quiz programmes such as *Take Your Pick* and *Double Your Money* became immensely popular. *The Ovaltineys* also returned to enthral a new generation of children.

Halfway through the decade Luxembourg's English service became mainly a music station. Numerous presenters fronted shows sponsored by record companies such as EMI, Decca and Capitol. The move to nightly music shows coincided with the introduction of the transistor radio and the emergence of youth culture and rock 'n' roll. Radio Luxembourg fully embraced the new musical phenomenon. Teenagers all over Europe would listen under the bedclothes to 'The Great 208', which offered a non-stop diet of popular tunes in marked contrast to the sparse offerings of the BBC. Such music on BBC radio was limited to a Sunday afternoon review of the current charts and a Saturday morning programme called *The Saturday Skiffle Club*, later renamed *Saturday Club* when the skiffle craze ended.

Listening to Radio Luxembourg, however, could be a frustrating experience for the dedicated music fan. The sponsored shows aimed to publicise as many of their new releases as possible, therefore the DJs never played a complete record and would usually fade it halfway through. Another problem was the famous 'Luxembourg fade'. Luxembourg's signal suffered from atmospheric interference and would often fade in and out. So if the DJs didn't fade your favourite record, the atmosphere would do it instead. Tuning to the station required endless patience and almost constant readjustment.

BBC radio continued to have some huge successes through the early part of the fifties. *Woman's Hour* is a magazine programme that began on the Light Programme in 1946 and still continues today. The programme contained reports, interviews and debates on health, education, cultural topics and short-run drama serials. Although aimed ostensibly at women, *Woman's Hour* also featured items of general interest. The programme switched to a morning slot in 1991 where it has remained ever since.

Alistair Cooke began his long-running series *Letter from America* on 24 March 1946. Originally born in Salford, Lancashire, Cooke relocated to America shortly before the outbreak of the Second World War. *Letter from America* was originally commissioned for only thirteen instalments, but finally came to an end fifty-eight years (2,869 episodes) later, in March 2004.

Down Your Way began on 29 December 1946. The programme visited towns around the United Kingdom, spoke to residents and played their choice of music. The programme vividly evoked the local and regional distinctiveness of Britain as it moved around the country. But while undoubtedly popular, the programme was not without its critics. Their view was that *Down Your Way* portrayed an increasingly old-fashioned and rose-tinted view of Britain. The critics' opinion was that the programme concentrated too much on market towns with pre-industrial roots and ignored industrial towns and urban conurbations. From 1987, until its demise in 1992, it was hosted by a different personality every week. This effectively turned it into 'Down My Way' as the guest presenters would visit a place of significance in their own lives.

Listen with Mother was a programme featuring stories, songs and nursery rhymes for children under five (and their mothers). It was broadcast on the Light Programme for fifteen minutes every weekday afternoon at 1.45pm, just before *Woman's Hour*. The theme music, which became synonymous with the programme, was the 'Berceuse' from Gabriel Fauré's *Dolly Suite for Piano Duet, Opus 56*. The programme's opening phrase 'Are you sitting comfortably? Then I'll begin,' etched itself into the British consciousness. At its peak, *Listen with Mother* had an audience of over a million. Like *Woman's Hour*, it was later transferred to the Home Service and lasted until 1982.

The BBC decided to try out an American radio format, the 'Soap Opera', although they called these programmes 'Serial Dramas'. *Mrs Dale's Diary* was the first major example of this type to appear on the BBC. The storyline featured a doctor's wife, Mrs Mary Dale, and her husband, Jim. Each episode began with a brief narrative spoken by Mrs Dale as if she were writing her diary. The programme began on the Light Programme on 5 January 1948, and subsequently transferred to the newly formed Radio Two in 1967, where it ran until 25 April 1969. A new episode was broadcast each weekday afternoon, with a repeat the following morning.

The Archers was first transmitted in the Midlands area as a pilot series on 29 May 1950. The BBC decided to commission the series for a longer national run. The series began on New Year's Day 1951 and continues to this day. In fact it is now the world's longest running soap opera. The programme was originally billed as 'an everyday story of country folk', but is now described by the BBC as 'contemporary drama in a rural setting'. *The Archers* is set in the fictional village of Ambridge in the equally fictional county of Borsetshire, which is supposedly situated between the real counties of Worcestershire and Warwickshire.

The BBC transmitted several other landmark broadcasts during the 1950s. The funeral of King George VI was broadcast on BBC Radio and Television on 15 February 1952. Nine days before, on the day the king died, all BBC programmes had been cancelled except for news bulletins and essential shipping forecasts.

The first ball-by-ball *Test Match Special* appeared on BBC radio on 30 June 1957. Live cricket had been broadcast since 1927, but originally it was thought that Test match cricket was too slow for ball-by-ball commentary to work. It was suggested that the Third Programme's frequency would be ideal for full ball-by-ball coverage, since at that time the Third Programme only broadcast in the evening. It soon became apparent that ball-by-ball commentary was compelling and *Test Match Special* has been heard on BBC Radio ever since.

The Goon Show

Undoubtedly the comedy hit of the 1950s was a programme that starred Peter Sellers, Harry Secombe, Spike Milligan and, initially, Michael Bentine. The first series, broadcast between May and September 1951, was titled *Crazy People*. After Bentine's departure at the end of the first series, the show was renamed *The Goon Show*.

The show's chief creator and main writer was Spike Milligan, whose scripts mixed preposterous storylines together with surreal humour, puns, catchphrases and peculiar sound effects. Many elements of the show satirised various aspects of life in Britain at the time. The talented cast complemented the bizarre scripts by utilising a wide range of increasingly wacky voices to portray the eccentric characters.

The Goon Show remained popular long after the show finished in 1960 and is continuously repeated at regular intervals. A special one-off edition of the show was broadcast as part of the BBC's fiftieth anniversary celebrations. Generations of subsequent British comedians cite the show as one of their main influences.

Hancock's Half-Hour

Hancock's Half-Hour was less fanciful than the Goons but equally influential. Comedian Tony Hancock was the star of the show. The series broke with the variety tradition, which was then dominant in British radio comedy. Instead of the normal mix of sketches, guest stars and musical interludes, the show's humour derived from characters and situations developed in a half-hour storyline. It could be said that *Hancock's Half Hour* was the first British situation comedy.

Hancock played an exaggerated version of his own character who lived at the dilapidated 23 Railway Cuttings in East Cheam. His sidekicks or, more often than not, his own embarrassing shortcomings, constantly thwart Hancock's fanciful ambitions in life. Sid James played Hancock's criminally inclined friend and Bill Kerr appeared as Hancock's dim-witted Australian lodger. From the third series, Hattie Jacques played Griselda Pugh, who was Hancock's secretary and Sid's occasional girlfriend. The show started on radio in 1954 and transferred to television two years later, so Hancock was the first British comedian to make the transition from radio to television.

Radio stations in the USA began experiments with broadcasting on VHF (very high frequency) using FM (frequency modulation) in January 1941. The first BBC VHF-FM radio transmissions started on 2 May 1955 from Wrotham in Kent. These transmissions offered better quality and were less susceptible to the interference often encountered on medium and longwave broadcasts. Medium wave transmissions frequently suffered interference from continental stations after dark. Longwave broadcasts experienced interference from unsuppressed electrical motors, thermostats etc.

This new service brought high-fidelity radio to around thirteen million potential listeners in London and the South East of England. The BBC gradually introduced more high-power transmitters followed by many low-power relay stations that filled in some significant pockets of poor reception. Development of the FM radio network was comparatively fast as most of the transmitters shared masts with BBC television.

All the initial FM transmissions were in mono. By 1961 there were twenty-seven FM transmitters in service and extra regional programmes were carried on FM only for the North East, East Anglia and South West regions.

The first stereo test transmissions began on 28 August 1962. The BBC used the Third Programme transmitter at Wrotham, which remained the only source of stereo radio for some years. It would take several decades for all the BBC's main networks and local stations to be broadcast in stereo.

The problem for the BBC was the distribution of stereo to its transmitters. At that time the organisation relied totally on the GPO's network of landlines for programme distribution and stereo required not one but two such lines. Also, these lines had to be carefully matched in terms of their quality and their length, and this proved to be difficult.

The BBC eventually solved the problem by using a system they called Pulse Code Modulation. This system could convert thirteen audio channels into one digital bit stream. This stream could be carried easily using the sort of link that carried television pictures. The BBC added stereo capability to Radios 2 and 4 in 1973 and stereo radio finally began to extend across the British Isles.

The BBC reorganised their three networks on 30 September 1957. The majority of the Home Service's lighter content transferred to the Light Programme. The Third Programme was renamed the BBC Third Network and cultural programmes were reduced from forty to twenty-four hours a week. The extra hours were used to incorporate the Home Service's adult education content and the Home and Light's sports coverage.

Despite its prominent position for nearly four decades, radio's popularity declined swiftly during the 1950s. The main reason for this was undoubtedly the ascendancy of radio's electronic progeny, television.

During the early part of the 1950s the BBC concentrated most of its efforts on radio. In 1950 there were twelve million radio-only licences and only 350,000 combined radio and TV licences. The BBC budget for television was a fraction of the radio one. However, the coronation of Queen Elizabeth II on 2 June 1953 was a turning point. This was the first time that a television audience exceeded the size of a radio audience. An estimated twenty million TV viewers saw the young Queen crowned. The following year saw the amount of combined sound and vision licences rise to over three million. The television age had arrived in Britain.

Chapter 8

Radio Down Under

In this chapter we look at the development of radio in the southern hemisphere. At the beginning of the twentieth century Australia was a newly federated country. The new government realised that the broadcasting spectrum should be regulated and wireless telegraphy quickly came under their control. The Wireless Telegraphy Act of 1905 was introduced and since then, broadcasting in Australia has remained the responsibility of federal governments. In the same year Marconi's company built Australia's first two-way wireless telegraphy station at Queenscliff in Victoria. Marconi and its main competitor, Telefunken, combined to form Amalgamated Wireless (Australasia) Ltd. in 1913.

That same year, the AWA established the Marconi Telefunken College of Telegraphy (later renamed the Marconi School of Wireless). This establishment made a valuable contribution during the two world wars. During times of conflict there is always a tremendous need for radio communications and Australia was able to maintain a high level of expertise. This has been attributed mainly to the effectiveness of the Marconi School of Wireless.

After the RMS *Titanic* tragedy in 1912, many nations around the world agreed to provide coastal communication services to ensure the safety of seafarers. The first Australian coastal radio station was established in Melbourne in 1912. More stations followed and by 1914 there were nineteen coastal stations around the Australian mainland. At first the stations communicated with Morse telegraphy.

Many people were attracted to the emerging technology. One such person was Ernest Fisk, who was born in Sunbury-on-Thames in 1886. Fisk joined the Marconi Company as a wireless operator in 1905 and visited Australia in that capacity in 1912. He returned the following year as Marconi's chief representative.

In July 1913 Telefunken and the Marconi Company pooled resources to form a new company called Amalgamated Wireless (Australasia) Limited. Fisk became a prominent employee. He was appointed managing director and destined to wield a powerful influence on all facets of wireless growth in Australia.

Ernest Fisk had long considered the possibility of direct radio broadcasts from the United Kingdom to Australia. On 22 September 1918 he received the first such message at his home in Sydney and then organised the first Australian radio broadcast on 19 August 1919. Fisk had arranged a lecture on the emerging technology to the Royal Society of New South Wales. He had arranged for a piece of music to be broadcast from one building to another at the end of the lecture. When he finished his talk he relayed the National Anthem and the patriotic audience rose to their feet. The lecture had been a triumph. He attempted to do the same thing at another public demonstration a few months later, but this one had a less satisfactory conclusion.

On this occasion Fisk was lecturing at Farmer's Blaxland Gallery. However, there was no telephone connection between his location and his headquarters so he told the engineer at Clarence Street to play the gramophone record continuously for an allotted period. Fisk finished his lecture and duly relayed the sound from his headquarters. The audience heard the National Anthem play and quickly rose to their feet. Unfortunately, the lecture had overrun. The engineer, thinking Fisk had probably finished his talk, flipped the record over. As the audience stood to attention they were suddenly treated to an orchestral version of *Yankee Doodle Dandy*.

After the First World War there were 900 radio amateurs in Australia and there were quite a few notable personalities involved at the dawn of Australasian broadcasting. Florence Violet McKenzie, known affectionately as 'Mrs Mac', was one of the region's most influential pioneers. She was Australia's first electrical engineer and a founder of the Women's Emergency Signalling Corps. She established a training establishment in Sydney, where some 12,000 pupils acquired Morse code and visual signalling skills. Mrs Mac was a lifelong supporter of technical education for women and fought successfully to have some of her female novices accepted into the all-male navy and, in so doing, helped form the Women's Royal Australian Naval Service.

She was also instrumental in setting up *Wireless Weekly,* which was Australia's first wireless magazine. It arose from conversations between McKenzie and another radio enthusiast, William J. MacLardy, who operated the amateur station 2HP. The first issue of *Wireless Weekly* had a print run of only a few hundred copies and went on sale on 4 August 1922. At first the magazine was exclusively for amateurs, but gradually became a broadcast listeners' journal. With the beginning of commercial broadcasting in 1923, it featured information about commercial stations and programmes and flourished, often exceeding sixty-four pages.

There are a few more radio pioneers worthy of note. Harry Kauper was the owner of the experimental station 5BG. His signal, using only a power of 5.5 watts, was heard as far away as New York and California, claiming a world record at the time. He started the Adelaide Radio Company in 1921, manufacturing and selling crystal sets, and assisted in the launch of a handful of radio stations in the pioneering days.

5AH was an experimental station established by Fred Williamson. It radiated from a transmitter at Kent Town every Tuesday for twenty minutes. The station was heard in New Zealand and America. Williamson was later to become a senior technician at the commercial version of 5AD.

Amalgamated Wireless wanted to sell sets sealed and tuned to only one frequency, and to supply the programming. By proposing a sealed set system, Fisk was trying to establish a monopoly. In his attempt to do this he gained quite a few influential friends. In September 1922 Fisk announced that one of the new directors of Amalgamated Wireless (Australasia) Ltd was Prime Minister William Morris Hughes. Having the prime minister as a director of a company was almost unprecedented.

Amalgamated Wireless also held numerous patents that other set builders would need to license from them, but many members of the industry had misgivings about the sealed set scheme. They suggested that the fixed tuning system made set design more difficult.

Fisk and the rest of the radio manufacturing industry lobbied heavily for the Australian government to introduce radio broadcasting. In May 1923, against objections by the Wireless Institute, which represented experimenters' views, Fisk's proposal was accepted and 'sealed set' broadcasting started. On payment of a fee, people received a radio that was set to the frequency of the stations they had subscribed to. A choice of stations would require a bigger payment.

The first major event supporting and promoting the emerging enterprise was held in the Sydney Town Hall in December 1923. The Wireless Exhibition was opened by Dr Earle Page, acting-prime minister, who indicated his consternation that the industry had been over commercialised to such a level as to limit general use. He declared, 'Wireless should be found in the working man's cottage as well as in the rich man's mansion.'

2FC in Sydney was the first to be licensed, on 1 July 1923, but its opponent 2SB was first to go to air. William J. MacLardy, the former editor of *Wireless Weekly*, was appointed as 2SB's station manager and he relocated the transmitter from his experimental station 2HP to a site in Phillip Street to commence initial test transmissions. The station had hoped to start broadcasting at the start of November but encountered a delay. They released this statement: 'Owing to the Commonwealth Government refusing to pass the Broadcast Receiving Sets [they] have been compelled to postpone the official opening of their service.'

The station eventually began broadcasting from the premises of the *Daily Guardian* newspaper on 23 November 1923. 2SB's first programme was a concert featuring a soprano, a bass, a contralto, a cellist, a baritone and a quartet. The baritone, George Saunders, was the station's first announcer. Saunders became an instant hit with younger listeners. He regularly read them bedtime stories assisted, or hindered, by 'Hector, the knowing bird' and 'Darkie the wireless horse'.

2FC was the next station to go on air, with its first test broadcast from its transmitter site at Willoughby on 5 December 1923, although they didn't begin regular programming until 9 January 1924. It appears that the reason for the delay was a problem in achieving the minimum transmission power. Part of the opening night's entertainment was a live relay of *The Southern Maid* from Her Majesty's Theatre. The station's call sign was derived from the initials of its owners, Farmer and Company, a Sydney department store. Ernest Fisk's company supplied 2FC's equipment and operated the station on their behalf.

After several months of transmission it was evident that listeners were confused by the similarly sounding call signs of 2SB and 2FC. So, in March 1924, 2SB changed its name and became 2BL. Melbourne became the next city to have two stations on air when 3AR began in January, with 3LO following in October 1924.

The listening public's response to the sealed set scheme was disappointing. Only 1,400 people took out licences in the first six months of 1924, but over 5,000 people had applied for an experimenter's licence. Admittedly, sets were scarce, but it was relatively easy to avoid the licence fee. This was possible by constructing your own set or modifying one you'd bought to receive more than one station.

Sealed sets were creating problems for the manufacturer, broadcaster and listener, with newspapers questioning their validity. This view was reinforced when the PMG announced that of the seventy-one receiving sets submitted for testing, only twenty-two had been approved. Therefore, an alternative scheme was sought.

The radio industry proposed an alternative two-tiered system: A and B licences. The Australian Federal Government accepted this compromise proposal on 17 July 1924. The 'A' licences were mostly financed by listeners' licence fees, imposed and collected by the government, while 'B'-class stations would have to generate their own revenue through advertising. This system was a combination of the British and American structures and Australia ended up with the parallel system of a national network alongside a commercial network.

The first 'A'-class stations were the original sealed set stations with the addition of 6WF in Perth, 4QG in Brisbane, 7ZL in Hobart, Tasmania and 5CL in Adelaide, which started on 20 November 1924. By the end of the year the number of listener licences was close to 40,000 and had doubled to 80,000 by the end of 1925.

6WF in Perth went on air on 4 June 1924. The station's name was taken from the initials of its owners, Westralian Farmers Ltd, who published the *West Australian Newspaper.* They originally broadcast on longwave believing that most of Western Australia would be covered. They eventually moved to medium wave on 1 Sep 1929 using 650 watts. Before moving, they conducted Australia's first stereo transmission; a concert with separate microphones linked to each of their transmitters. Listeners needed two radios to hear stereo.

5CL started transmitting to Adelaide on 20 November 1924, only two days after being granted a licence by the Postmaster General's office. Their manager, W. Smallcombe, was a man of many parts. Not only was he the chief announcer but he also sang and played the piano on air. For their distinctive hourly time signal they would relay the chimes of the local post office clock using a remote microphone.

7ZL, owned by the Associated Radio Company of Australia, started in December 1924, in just one room in *The Mercury* newspaper office room before moving to the old Hobart railway station in 1928.

And 4QG, owned by the Queensland government, began in July 1925 with a licence to transmit on 779 kilohertz at 5,000 watts. The station was officially opened by the premier, William Gillies. Unfortunately, their debut broadcast from the Tivoli Theatre Orchestra had to be abandoned due to technical difficulties. Their opening broadcast was described by *The Queenslander* newspaper as 'generally disappointing'.

4QG soon mastered the art, though, and regularly broadcast from various locations. In September 1927 the station aired a mammoth seventy-one outside broadcasts. They made plans to cover the arrival of aviator Amy Johnson to Brisbane Eagle on 30 May 1930. However, this turned out to be a very short outside broadcast. While approaching Eagle Farm Airport her plane struck a fence, bounced over and crashed into a cornfield. All the announcer had time to say was 'Christ, she's crashed,' and the broadcast was immediately cut short. Luckily, on this occasion, Amy emerged from the wreckage unhurt, smiling and waving to the crowd.

The Postmaster General's department announced on 15 November 1924 that nine 'B'-class broadcast licences had been decided. Of those nine licences, only five actually made it to air. The first 'B'-class station to receive a licence was 2BE, owned by the Burgin Electric Company. According to station manager Oswald Mingay, 2BE started transmissions in November 1924. However, information has been found in the National Archives of Australia that seems to suggest they were still preparing to broadcast during 1925.

The first 'B'-class station to officially start broadcasting was 2UE, which went on air on Australia Day 1925. This was followed the day after by 2HD in Newcastle on 27 January 1925. South Australia's first 'B'-class station was 5DN, which went live on 24 February 1925.

2UE was managed by Cecil 'Pa' Stevenson, who operated the station from his house in Maroubra using home-made equipment. The studio was his dining room, the transmitter was situated on his veranda and the 24-m-high towers were in his backyard. Their original broadcasting hours were 8pm to 10pm every night. Cecil would frequently whistle while swapping the records over to let listeners know that he was still on air. 2UE eventually moved to premises in

Stevenson's 'Radio House' shop and was the first station to experiment in sending still pictures by radio.

2HD was launched by Harry Douglas with only twelve records in the library. He used the equipment from his experimental stations 2XY and 2CM, transmitting with a power of 10 watts from a room above his tyre shop in Hamilton.

5DN was owned by the Adelaide Radio Company and the owners of the Hume Pipe Company. Stella Hume was a regular announcer as the studio was based in her Parkside home. 5DN's first programmes were classical music broadcasts from the Elder Conservatorium and lectures from Adelaide University. 5DN commenced daily transmissions in December 1926 using 500 watts. On 12 August 1927 it introduced a programme called *Super-Het* to answer listeners' questions regarding technical problems with their receivers.

Amateur broadcasters were allowed to continue operating in the longwave and shortwave bands. In Melbourne, they were also permitted to broadcast on the medium wave band on Sundays between 12.30 and 2.30pm. All commercial stations were required to close down during this period.

The British government nationalised radio in 1926 by buying out the British Broadcasting Company and forming the British Broadcasting Corporation. That same year, the Australian government held a Royal Commission into wireless broadcasting. The government didn't immediately follow the British lead but they did encourage the 'A'-class stations to amalgamate in order to maximise efficiencies and maintain standards.

The 1927 Royal Commission recommended the licence fees be shared and that the 'A'-class stations should cooperate to provide better services and wider coverage. The tactic would standardise the service across the country, with larger capital city stations effectively supporting the smaller country town stations. Naturally, the larger stations refused to agree to this.

In 1928, to break the impasse, the government established the National Broadcasting Service to provide the service and coverage the existing stations were unwilling to provide. As the licences for the 'A' stations came up for renewal they were revoked and issued to the National Broadcasting Service. The government would purchase the ousted station's transmitters and studio equipment for the new public service.

The Postmaster General's department was given the responsibility of running the new service.

In 1929 the Australian government nationalised the transmission facilities and granted a three-year contract to the Australian Broadcasting Company, which consisted of Fullers Theatres, Greater Union Theatres and music publishers J. Albert & Son. They were to take over all 'A'-class stations and produce programmes on a national basis. Originally ABC was to be allowed to broadcast advertisements but this was dropped from the final bill. Instead, radio listeners' licences were chosen to fund it. Licence fees for radio and TV were finally dropped in the seventies and ABC's money now comes from federal government appropriation.

It gradually became politically difficult for the Postmaster General's office to sustain the running of the National Broadcasting Service so the government came up with a solution. When their contract ran out in 1932, the Australian Broadcasting Company was nationalised and the Australian Broadcasting Commission was established. This meant that Australia had twelve stations run by the ABC and forty-three commercial stations.

Australia was a leader in the use of shortwave broadcasting to transmit overseas. In 1927 Amalgamated Wireless (Australasia) conducted a series of transmissions to Britain. These regular broadcasts were heralded by a kookaburra's laugh. Radio Australia was formally incorporated as part of the ABC in 1939.

The Federation of Australian Radio Broadcasters was established in 1930. This was an industry body with a remit to look after and promote commercial radio interests. In 1934 the commercial radio sector pulled off a great scoop when it won the rights to broadcast commentary of cricket matches between England and Australia. The Test Match series was being played in the United Kingdom that year and this greatly boosted the amount of radios and licences purchased.

In New Zealand radio licensing was first introduced in 1923. Stations sprang up in all parts of the country, including 1YA in Auckland, 2YB in Wellington and 3AC in Christchurch. Radio manufacturers and shops selling musical instruments and sheet music ran many of these stations, which were often called 'trade' stations. Overseas radio manufacturers often supported these local businesses.

Amateur operators and radio societies also broadcast entertainment programmes regularly. The Radio Broadcasting Company of New

Zealand was established as a national operation in 1925. This was after an earlier plan by the NZ Co-Operative Dairy Company to operate a station aimed at dairy farmers was rejected.

Call signs were allocated by the government and divided New Zealand into four radio districts. The suffix 1 was allocated to Auckland (roughly a line between King Country, Taupo, Whakatane and all points north), Wellington and the rest of the North Island plus Nelson. Marlborough and the Chatham Islands were allotted call signs beginning with 2, while the call sign 3 was assigned to Christchurch (Canterbury as far south as South Canterbury, plus Buller and Westland). The suffix 4 was assigned to Dunedin and the rest of the South Island.

The year 1931 would prove to be pivotal in the early history of radio broadcasting in New Zealand. That year royalty disagreements shut down numerous stations and the Napier earthquake knocked several more stations off air. Yet the most important change was the expiration of the Radio Broadcasting Company's agreement. It was not renewed and the government established the New Zealand Broadcasting Board instead.

The NZBC pressed ahead with its own extension of radio coverage. High-powered transmitters were ordered for use in Auckland, Wellington, Christchurch, and Dunedin and were sited at various points to achieve best coverage. New Zealand's main broadcasting problem was reaching the relatively small pockets of population concentrated in the fertile areas that lay among mountain ranges. The board decided to give financial aid to a number of private broadcasting stations operating in areas where reception of its stations was unsatisfactory.

The New Zealand Broadcasting Board's tenure, however, was brief. In 1935 the country had gone to the polls and resolutely returned its first Labour government. One of the first legislative measures passed was the Broadcasting Act 1936, which came into force on 1 July. One of the main provisions of the legislation was the abolition of the New Zealand Broadcasting Board. All the rights, property, liabilities, and engagements of the board were transferred to the Crown. As a result, for the foreseeable future, broadcasting was to be administered as a government department.

Undoubtedly the most controversial aspect of the Broadcasting Act was the prohibition of advertising, except by commercial stations controlled by the government. Deprived of their revenue stream, all privately owned commercial stations were forced to close down.

Despite the government's virtual broadcasting monopoly, listeners in New Zealand could hear a wide variety of stations from other sources. At night stations from Australia, Hawaii, California and even Asia could be easily heard, as there were hardly any local stations to listen to. Stations such as 2FC in Sydney, KFI in Los Angeles and KGU in Honolulu became very popular.

In Australia the expansion of radio continued rapidly. By the early 1940s there were about 130 commercial stations and a similar number of ABC stations available. The ABC had national commitments including news, education, parliamentary broadcasting and culture. The commercial stations were much more local and community-orientated in nature.

On 2 February 1942 Australia's first nationally sponsored morning serial was broadcast across a network of stations. *Big Sister* originated from 2UW in Sydney, although the scripts were American. The show, sponsored by Lever and Kitchen, was heard five mornings a week. Throughout its five-year run it was the top-rated daytime show and was the forerunner of many other daytime serials in that genre.

The years after the Second World War are considered to be radio's golden era. Radio drama became very popular in Australia. The most popular dramas were *Blue Hills, The Lawsons* and *Dad and Dave*. Australian children had their own radio programme called *The Argonauts Club,* which ran for thirty years from 1941 to 1971. Over 50,000 Australian children became members of the club during that era.

Many American formats were adapted for Australian radio. *The Lux Radio Theatre* began in Australia on 19 March 1939 after five successful years on American radio. Other American shows adapted for Australia included *Inner Sanctum Mysteries, The Falcon, Dragnet, The Witches' Tale* and *Nightbeat.*

The Broadcasting & Television Act had originally been put into practice in 1942 but a major amendment was made in 1948. This was to introduce a regulatory body, the Australian Broadcasting Control Board. The new authority maintained there was no room for new stations on the AM band so experimental FM broadcasts began.

There was another major amendment in 1956 to introduce television and a further inquiry into FM in 1957. However, the commercial operators were unwilling to invest in the new infrastructure that would be required. Planned FM services were shelved and eventually the Australian

Broadcasting Control Board authorised the use of the international VHF FM band for television in 1961.

The closure of the FM experimental stations, together with the Australian Broadcasting Control Board's recalcitrant position on AM broadcasts, meant that no new competition came onto the scene. There was widespread dissatisfaction with the government for not introducing FM quality broadcasting. This emanated mainly from people who wanted to hear 'fine music' (classical or jazz) on the airwaves.

In 1961 Dr Neil Runcie formed the Listener's Society of New South Wales, whose major intention was to establish subscriber-supported FM music stations. This concept had had some success in the USA with the Pacifica stations and a few educational FM stations.

In the same year the University of New South Wales was granted a licence to broadcast using an old RAAF transmitter on 1750 kilohertz. This station was assigned the call letters VL-2UV and their licence was for university lectures only with no music allowed. By 1962 they were transmitting thirty separate courses for over 1,000 fee-paying students. They even experimented with television programmes on UHF in 1966.

These two organisations were the progenitors of a movement to provide more diversity in Australia's radio broadcasting. Ultimately this movement led to the establishment of the third tier of broadcasting in Australia, the public or community sector.

The third prong of this movement came from Australia's ethnic communities. Australia had undertaken the biggest programme of immigration in the world after the Second World War and the country's population had almost doubled in twenty years. By the late sixties this large group of immigrants, many of them from non-English speaking backgrounds, was reaching political maturity. Ethnic communities were demanding a more open media.

In 1964 the Australian Broadcasting Control Board had allowed for up to ten per cent of broadcasting time to be in non-English languages. The commercial sector utilised this provision for revenue and stations such as 2CH and 3XY sublet airtime to ethnic groups. However, as commercial radio became more competitive and format-driven, the amount of ethnic broadcasting decreased until, in 1972, there were only thirty-six hours of ethnic broadcasting in the country in six languages.

The fourth group seeking access to the airwaves was made up of young political activists. There was a climate of political unrest in the late sixties with people protesting against Australia's involvement in the Vietnam War. In 1971 Melbourne University students put together a pirate radio station in the Union building and broadcast anti-government messages. It was only on air for a few hours before Federal Police broke in and confiscated the transmitting equipment. That same year the weak response to the Springbok rugby tour demonstrations led Brisbane students to look at forming their own radio station. This eventually became the alternative music station 4ZZZ.

In the early part of 1972 the Australian Broadcasting Control Board held another broadcasting inquiry. One of the proposals was the abolition of licence fees for radio and TV. The Whitlam government finally abrogated them in 1974 and ABC's funding now comes from federal government appropriation.

Another of the recommendations of this report was the introduction of public access broadcasting. This differed from the other two radio sectors because of community involvement in both the management and programming of the station. The 'community' could be a geographically defined district or a society of special interest. These stations were to be non-profit and community owned. They wouldn't receive government funding and were only allowed limited advertising.

As mentioned earlier, the ethnic communities of Australia were pushing for access to the airwaves throughout the early seventies. This lobbying assisted in the implementation of community broadcasting in the mid-seventies. There are five full-time ethnic community radio stations and about forty-five others broadcasting some ethnic programming.

However, while this process was unfolding, a number of other approaches were also tried. The ABC had been persuaded by Gough Whitlam's government to open two 'open access' stations in 1975. These stations, 3ZZ in Melbourne and 2JJ in Sydney, rapidly became de facto ethnic broadcasting stations.

Also in 1975 two experimental stations were opened in Sydney and Melbourne to broadcast information to ethnic communities about Medicare. These temporary stations, 2EA and 3EA, were allowed to stay on air. Incidentally, the EA in the station's call signs stood for ethnic affairs. The Malcolm Fraser government eventually set up the Special Broadcasting Service (SBS) to run the stations when the ABC showed reluctance to take them on board as part of its charter renewal.

The other main initiative from the 1972 inquiry by the ABCB was the introduction of FM broadcasting, although this was planned for the UHF band rather than the internationally used VHF band. This was suggested by the radio manufacturing industry who wanted to sell sets that were only usable in Australia.

By the time the recommendations were implemented in the mid-seventies, common sense had prevailed. The UHF proposals had been rejected in favour of the more popular VHF band. The first use of FM in Australia was for non-profit community-based public broadcasting. The first station on FM was 2MBS in Sydney, which started in the latter part of 1974; 3MBS in Melbourne followed the following summer. Both stations had a 'fine music' format. The ABC entered the medium in 1976 with the establishment of ABC-FM based in Adelaide.

VL-5UV in Adelaide is often cited as Australia's first public access station. The station, which made its first broadcast in June 1972, was established as a direct educational outreach of the University of Adelaide. At this stage its wavelength was off the AM band due to legal requirements and was restricted to twelve broadcast hours per week. VL-5UV couldn't really be described as a public broadcaster at this point as it didn't have community access or ethnic programmes. Its programming was strictly didactic; lectures were recorded and rebroadcast later.

VL-5UV wanted to expand the station's programming and move to the access and participation model that was to eventually characterise public broadcasting. However, they were not licensed to do so and had to wait for legislation that would permit them to do this.

The Wireless & Telegraph Act was eventually introduced in 1974. The new legislation meant that that VL-5UV could finally move into the AM band. They switched to 530 kilohertz, which was later upgraded to 531 kilohertz. The station's name was shortened to 5UV at this point and many specialist and ethnic groups were given airtime.

5UV gained an FM outlet in October 2001 and AM broadcasting ceased the following January. Gradually the station became more independent of the university and is now located in its own premises. The station adopted a two-part name change process: Radio 5UV is now known as Radio Adelaide, following a provisional period known as 5UV Radio Adelaide.

The commercial sector had been unwilling to commit to FM broadcasts when the spectrum was first offered in the early 1970s, so the FM spectrum went to public and community broadcasters instead. The commercial operators swiftly realised the error of their ways and began to petition the Australian government.

In 1980 the government offered a limited number of FM licences: two in Melbourne and Sydney and one in the other state capitals, the same as in 1924 when the 'A'-class licences were first introduced. However, these licences went to new companies rather than the existing stations. The new FM commercial stations swiftly became profitable and became the ratings leader in most markets.

After much lobbying, the government allowed a chosen few AM stations to convert to FM. They instigated an auction of the FM frequencies and the resultant bidding war to win the right to convert severely affected the economies of the commercial sector. In addition, the radio industry became ensnared in the media buying frenzy that accompanied the widespread entrepreneurial boom during the latter part of the eighties. As a result, many stations got into financial difficulties and changed hands.

A good example of a station changing hands was 3EON in Melbourne. It was the first commercial FM radio station in Australia, beating Fox FM to the title by two weeks. Eon FM's inaugural FM broadcast took place on 11 July 1980. At first the station had no playlist and deliberately avoided Top Forty songs, widely publicising that they would feature songs that 'would not be played elsewhere.'

Although the station out-performed other FM stations, the AM stations comprehensively beat it. In 1981 Eon FM abandoned the album rock format and began playing Top Forty records instead. This changed the station's fortunes and it eventually topped the ratings in 1985. 3EON FM was sold to the Triple M network the following year and eventually rebranded to 3MMM on 27 November 1988.

As the 1980s became the 1990s, large networks purchased many stations. In turn other predatory networks would swallow them up: 1992 saw a monopolistic arrangement take place whereby the Austereo network purchased the Triple M network owned by the Hoyts Group, then the Village Roadshow Media Company purchased Austereo. This deal was unpopular due to the fierce rivalry between the two radio networks. Village Roadshow and Hoyts were also direct competitors in the film industry.

In March 2011 Southern Cross Media launched a takeover bid of the Austereo Group. The following month shareholders of the Austereo Group accepted the takeover bid, giving Southern Cross Media a ninety per cent share in the company. Southern Cross Media and Austereo officially merged in July 2011 to form Southern Cross Austereo.

The new group now has two distinct radio networks amongst its media portfolio. The Hit Network is a focused popular music format targeted at the eighteen to thirty-nine age group. The other brand is a bit more complicated depending on where the station is situated. Triple M is effectively comprised of four networks: the longest-running and 'main' version is the Metropolitan network, which focuses on a mix of rock, sport and comedy, and there are four of these stations all based in capital cities. The second is the digital radio network, which comprises all the rock, sport and comedy stations along with specialist brands such as classic rock and country. The third is the Greatest Hits network, which plays a mix of oldies depending on local audience demographics. The fourth and final network, the Classic Hits stations, are similar in style to the Greatest Hits outlets but tend to favour more golden oldies from the 1950s and 1960s.

There are approximately 300 commercial radio broadcasting stations currently operating in Australia. Large networks also own many of these stations. These include Grant Broadcasters, which is a regional radio network that also has a small number of metropolitan radio stations. It currently has over fifty stations in its portfolio. The Broadcast Operations Group operates forty-two radio stations. Almost half of these form its AM network and feature local news, music and syndicated talk programmes. Their FM network broadcasts adult contemporary music.

There are several smaller networks too. Ace Radio has thirteen stations, mostly in Victoria State. The Australian Radio Network owns the Classic Hits and Mix FM stations while Nova Entertainment possesses seven Nova and Smooth FM stations. The Capital Radio Network owns seven stations in rural New South Wales

Australia has had an extensive commercial radio network since the 1920s. Though, as we've heard previously, radio in New Zealand remained firmly under the aegis of the government. The National Broadcasting Service, and its successor – the New Zealand Broadcasting Corporation – had been operating since 1936. No privately owned station had been allowed to operate for over thirty years. In the mid-1960s the impetus for privately owned commercial radio came from a rather surprising area.

The genesis of modern-day commercial radio in New Zealand took the form of a group of radio pirates in international waters off the coast of Auckland. Radio Hauraki transmitted from studios aboard the ships

Tiri and its successor, *Tiri II*. The station officially started broadcasting on 4 December 1966 and its Americanised Top Forty music format was immensely successful. Radio Hauraki's success forced the government to create the New Zealand Broadcasting Authority in 1968. This authority issued the first two private commercial broadcasting licences on 24 March 1970. One of the successful applicants was Radio Hauraki.

Unfortunately for the proponents of commercial radio, the allocation of licences was slower than expected. By 1972 only five private stations were on the air in New Zealand. In response to public pressure, the 1974 Labour government pushed through legislation that split the state-run New Zealand Broadcasting Corporation into three distinct sectors. One of these was Radio New Zealand, an amalgamation of commercial and non-commercial stations and networks. At this point more commercial stations were introduced into markets around New Zealand.

In 1981 new legislation allowed the newly formed Broadcasting Tribunal to issue FM broadcasting warrants for the first time. This led to twenty-two private and thirty-seven government-owned commercial radio stations in 1984. Moreover, it was the introduction of the Radio Broadcasting Act in 1989 that generated the biggest expansion. All available frequencies around the country went up for tender and new ownership groups were formed. Commercial radio in New Zealand became arguably the most deregulated in the world. By April 1993 over 200 new frequencies were active and the increase continued exponentially. In 2004 over 700 frequencies were available for broadcast on AM and FM in New Zealand. This was possibly the largest number per capita anywhere in the world.

The reforms in radio structure were not restricted to the commercial sector as the government reviewed its own assets. 'Ruthanasia', a combination of 'Ruth' and 'euthanasia', was the deprecatory appellation given to the period of free-market economic reform conducted by the New Zealand government during the latter half of the 1990s. The 'Ruth' in Ruthanasia referred to the then minister of finance, Ruth Richardson. As a result of this reform, the government's commercial radio operations were sold to a conglomerate for $89 million in 1996. The Radio Network (TRN) was a partnership between Independent News and Media and Clear Channel.

Other private ownership groups were quick to react to the formation of this media giant. In early 1997 Energy Enterprises merged with Radio

Pacific and Canadian-based CanWest purchased the Frader Group. In May 1999 the Radio Pacific-Energy Enterprises group completed a takeover of Radio Otago and evolved into RadioWorks Ltd. Also, in May 2000, CanWest announced the successful purchase of RadioWorks and the commercial radio battle was down to just two major competitors: the Radio Network and CanWest Global. Both groups continued their acquisition spree and by 2005 TRN and CanWest owned or controlled over 350 frequencies in New Zealand.

These mergers and acquisitions would become a regular occurrence elsewhere in the world and, despite the consolidation, the radio industry within the Australasia region remains vibrant and healthy. The Australian commercial radio sector has been very progressive in its acceptance of new technology in recent times. These radio stations are morphing into multimedia companies and a natural extension for them is to actually become retailers of music. Listeners to Nova Entertainment stations can buy and download songs or albums featured on their playlists. Australia Radio Network's station, The Edge 96.1, provided a service that allowed listeners to find out the name of a song playing on the station and download it as a mobile phone ringtone.

The emergence of new transmission technologies sounded the death knell for older and more traditional methods. On 31 January 2017 the international shortwave transmitters at Shepparton in Victoria were switched off and Radio Australia was silenced. The three companion transmitters in the Northern Territory were also closed, thus ending the domestic Outback Service for isolated communities. This angered cattle station owners, indigenous ranger groups and fishermen, who complained it was done without community consultation. They maintained it would deprive people in these remote areas of vital emergency warnings.

Radio Australia's demise didn't mean the end of shortwave broadcasting on the continent. 4KZ maintain a shortwave relay of their medium wave station with the same call sign. The shortwave transmission serves remote areas of North Queensland. Unique Radio currently broadcasts from Gunnedah in New South Wales using amateur radio equipment, and there are also plans for another shortwave station to start broadcasting to the state of Victoria in the near future.

The move to abandon shortwave coincided with ABC's decision to expand its digital content offerings, including digital radio, online and mobile services, together with FM services for international audiences. They maintained that the money saved through decommissioning the shortwave service would be reinvested in a more robust FM and digital transmitter network.

A staged roll-out of commercial digital services began with Perth and Melbourne in May 2009 with Adelaide, Brisbane and Sydney soon following suit. The ABC national service was delayed due to funding issues but eventually started in July. So far eight major conurbations have a national and local digital service. The federal government allocated funding to help community broadcasters begin digital broadcasting and these stations were guaranteed spectrum on each of the local services. The first of these city-wide licences began in 2010 and spread to various regions the following year.

Digital broadcasts in Australia utilise the DAB+ standard and are available via multiplexes in Melbourne, Adelaide, Sydney, Brisbane, Perth, Canberra, Hobart and Darwin. The national government-owned networks contain a combination of ABC and SBS services, together with the commercial radio stations in each market. These commercial offerings are chiefly simulcasts of their AM or FM content. The ABC also have a number of digital-only radio stations such as ABC Country, Double J, ABC Jazz and a special events station, ABC Extra, which is used to provide additional coverage for special events.

Chapter 9

Radio Fights Back

In Britain, television really began to catch on when the first commercial station began on 22 September 1955. The new Independent Television Authority (ITA) began its broadcasts with live coverage of an opening ceremony and banquet at the Guildhall in London. After the Guildhall banquet, the main programmes got under way. They included half an hour of drama excerpts, news bulletins, a variety show and a boxing match.

The first commercial came a little more than an hour into the schedule. Viewers saw a tube of Gibbs SR toothpaste in a block of ice, with a voice over pronouncing it 'a tingling fresh toothpaste for teeth and gums'. There were another twenty-three advertisements during the evening, promoting products ranging from Cadbury's chocolate to Esso petrol.

Yet BBC Radio staged an effective spoiling tactic that managed to upstage the launch of ITV. The BBC broadcast a controversial episode of *The Archers* featuring the death of Grace Archer, a leading character in the serial. The episode pulled in an audience of twenty million and generated a lot of publicity in the newspapers. The BBC has always maintained that this was coincidental and not deliberately planned to divert attention from the opening night of commercial television.

The arrival of commercial television created great controversy. Some compared its arrival to that of the great plague. Winston Churchill was unimpressed and described it as 'a tuppenny Punch and Judy show'. The BBC, hindered by its dependency on a licence fee fixed by the government, saw its audience share drop to twenty-eight per cent. Nevertheless, competition gave the BBC a necessary jolt, forcing it to revamp its drama and news presentations.

In America, radio also played second fiddle to television. By 1950 more than four million television sets had been sold and the radio set had lost its valued place in American living rooms to the new arrival.

Radio stations found it difficult to retain staff as sales people and programme personnel defected en masse to the new medium.

Perhaps the biggest threat posed to American radio by television was financial. Advertisers were seduced by the newer, sexier medium and gradually reduced their radio budget. Radio station owners simply adjusted their tactics to cope with the new economic realities. Most local stations couldn't afford to finance large variety shows or dramas so they turned to a cheaper alternative instead. They broadcast pre-recorded music linked by disc jockeys.

Many station owners realised that the days of broadcasting to a substantial number of listeners were over and elected to aim their transmissions at a niche audience instead. By 'narrowcasting', they identified target audiences and created formats to specifically appeal to those listeners. By the end of the decade, various musical formats such as rock 'n' roll, jazz, classical and country had proliferated.

A strong local identity was essential and many stations seized every opportunity to get involved in local affairs. They increased local news bulletins and frequently publicised events in their area. This localness, combined with specialist programming, helped to save American radio from its television-induced decline.

The new style of American radio featuring music laced together by a fast-talking DJ became extremely popular. The success of this new broadcast format was boosted by the emergence of rock 'n' roll. As older listeners transferred their allegiance to television, youngsters replaced them as the core radio audience.

Listeners in Europe had to wait a while before they were able to experience the non-stop diet of pop music that proliferated in America. The BBC only had a limited amount of suitable programmes each week and Luxembourg's schedule was full of sponsored record company shows.

In 1958 a group of pioneers circumvented European broadcasting regulations by transmitting from ships or fixed maritime structures in international waters. These offshore stations, or 'pirates' as they became known, were able to exploit a loophole in maritime law.

A country's jurisdiction only extended as far as territorial waters. In most cases this was 5km out from the coast. In international waters a vessel need only recognise the laws of the country whose flag it sailed under; if the law of the flag state had no objection to international marine broadcasting then the ship could broadcast quite freely and legally.

The idea was not new by any means, as this had been done twenty-five years earlier in May 1933. The first 'pirate' offshore radio station was broadcast from the SS *City of Panama*, a floating gambling casino off the shores of California. RKXR was permitted to transmit non-commercial programmes on 815 kilohertz with a power of 500 watts. Yet, when the ship started broadcasting it had an output power of 5,000 watts and pumped out popular music and commercials. The salt-water pathway ensured a very strong signal in Los Angeles but caused severe interference to legally licensed land-based stations. RKXR was a hit with listeners and advertisers but the positive reaction to the station caused a flurry of diplomatic activity and action was taken to close it. By August the ship was towed into Los Angeles harbour and no more was heard of the station.

A few stations broadcast from man-made structures such as old army and navy forts in the Thames estuary. Radio and TV Noordzee, a Dutch pirate, even went as far as building their own platform known as REM Island.

Europe's first 'offshore' station was a Danish pirate called Radio Mercur, which started regular transmissions on 2 August 1958. Other ships soon joined the station and during its heyday in the mid-1960s, there were at least a dozen similar operations pumping out music from international waters in the North Sea.

Radio Caroline became the first British pirate when it anchored off Felixstowe and started broadcasting from the MV *Caroline* in Easter 1964. It was soon joined by another station called Radio Atlanta, which used a ship called the MV *MI Amigo* as its transmission base. The two stations announced a merger in July 1964. Radio Atlanta was renamed Radio Caroline North and set sail to the Isle of Man while the station aboard the MV *Caroline* became Radio Caroline South.

Other ships soon joined the broadcasting fleet. Radio Scotland broadcast from a former lightship called *The Comet.* Radio 270's home was the *Oceaan 7* anchored off Bridlington and Scarborough. The MV *Olga Patricia* (later renamed the MV *Laissez Faire*) was unique in that it housed two pirate stations: Britain Radio, which broadcast easy listening music, and Swinging Radio England, which was a fast-moving pop music station staffed mostly by American DJs. The two stations were owned by a group of businessmen from Texas, but the brash American style proved unpalatable for British listeners and failed to make money. Swinging Radio England was closed down after a turbulent six months on air.

Undoubtedly the most successful pirate of the period was Radio London, which transmitted from the MV *Galaxy* off the coast from Frinton-on-Sea in Essex. Like Swinging Radio England, Radio London had US backing, but they adapted the slick, fast-moving American style and made it more acceptable to British ears.

Other stations such as Radio Sutch, BBMS, Radio Invicta, Radio 390, Radio Essex and Radio City used the abandoned military and naval forts in the Thames estuary as their broadcasting base. Broadcasting from these man-made platforms proved problematic for the various stations involved. Radio City and Radio 390 faced continual legal action from the GPO who insisted that they were illegally broadcasting from within British territorial waters.

Stations like Radio Veronica in the Netherlands and Radio London in Britain proved immensely popular with the listeners. They were only silenced when the various European governments introduced legislation to outlaw them. These laws made it illegal for any citizen to work for, supply goods to, or, more importantly, advertise on any such station.

A financial dispute, which led to the death of Radio City owner Reg Calvert, was the catalyst that finally prompted the British government to act. The Marine Offences Act was introduced on Monday, 14 August 1967. Most stations could not survive without advertising revenue and reluctantly complied with the law. Radio Caroline, however, remained defiant. The two Caroline stations flouted the law and carried on for another year until a financial dispute with the tender company forced them off air. In the early hours of 2 March 1968 the two ships were boarded and seized by representatives of the Wijsmuller Transport Company and towed back to Holland.

The British government realised that outlawing the offshore stations would be hugely unpopular, so they persuaded the BBC to introduce a replacement. On 30 June 1967 Edward Short, the then Postmaster General, announced in Parliament that the BBC would open their new 'pop channel' in the autumn. A month later, BBC Director of Radio, Frank Gillard, announced plans to 'kill off' the Light Programme, Home Service and Third Programme. In future it would be 'radio by numbers'.

Radio One, the new pop music network, was launched on the morning of Saturday, 30 September 1967. The other networks also received a revamp. The Light Programme became Radio Two and the classical Third Programme was renamed Radio Three. The oldest channel, the speech-based Home service, became Radio Four.

A few minutes before 7am, Radio One's theme was played. *Theme One* was especially composed for the launch by George Martin, the Beatles' producer. The first DJ was Tony Blackburn, who presented the new show *Daily Disc Delivery*. Blackburn fired the first jingle and played his theme tune, which was the same one he had used on the pirate stations. As the tune *Beefeaters* by Johnny Dankworth faded he made his first announcement and the station was on the air.

The first record played was *Flowers in the Rain* by The Move, which was the number two record in that week's Top Twenty. The current number one record by Englebert Humperdink would not have fitted the station's sound. The second single was *Massachusetts* by The Bee Gees and they became the first live group to appear on the new network when they were guests on *Saturday Club*, at 10am.

There was a pleasant surprise for fans of pirate radio. The new station had 'adapted' Radio London's jingles to suit their own needs. Therefore many listeners were familiar with the sonic idents that interspersed with the pop records. One of the reasons jingles from PAMS were used was that the Musicians' Union would not agree to a 'one-off' fee for the musicians and singers if the jingles were made 'in-house' by the BBC. The union had wanted repeat fees each time one was played.

There were problems, however. Radio One was the only national BBC network that didn't get an FM outlet. The official reason for this exclusion was that there was not enough space on the FM dial, although there were no licensed independent commercial stations at the time. Radio One's transmissions used the sixteen existing medium wave relay transmitters of the Light Programme. Most were of comparatively low power so many parts of the country couldn't get a good signal. Scotland, Devon, Cornwall and several parts of Wales encountered severe reception problems after dark.

Many youngsters, the station's target audience, were dissatisfied about the replacement for the offshore radio stations. This was despite the fact that the majority of Radio One presenters had been recruited from the pirates. Of the twenty-nine presenters listed on the opening schedules, only twelve had a BBC background. However, the very fact that the newcomer was part of an establishment institution such as the BBC was a turn-off for some. The number of DJs hired at the launch far exceeded the positions available and eventually the numbers were whittled down.

The other problem was quite a large one for a network that was publicised as a 'non-stop pop channel'. The BBC had to observe 'needle time' regulations imposed by the Phonographic Performance Society. The corporation were only allowed to play 'commercial gramophone records' for seven hours per day over both Radio One and Two. Therefore a lot of programmes had to be shared between the networks. In fact, on the first day only five and a half hours of programmes were broadcast on Radio One alone.

Despite this, the new station had doubled the Light Programme's audience within the first month of launch. It became the most listened-to station in the world, with audiences of over ten million claimed for some of its shows. The station undoubtedly benefited from the lack of competition, as Independent Local Radio didn't exist for the first five years of its existence.

The Radio One publicity department made sure that their presenters were rarely out of the newspapers. 'Personality' DJs like Noel Edmonds, David Hamilton and Dave Lee Travis became household names. The touring summer live broadcasts called the *Radio One Roadshow* drew some of the largest crowds of the decade and the station undoubtedly played a role in maintaining the high sales of single records.

Radio One received the lion's share of publicity leaving the Light Programme's replacement to its own devices. Radio Two carried much of its predecessor's output but some of the old Light programme favourites were lost. Long-running shows such as *Music While You Work* and Sunday morning's pop show *Easy Beat* were all casualties of the shake-up.

Radio Two kept their longwave and FM outlets, with new medium wave transmitters opening on 1484 kilohertz in Scotland to replace those lost to Radio One. The network settled down as an easy listening station playing gentle pop and oldies with specialist programmes covering musical genres such as country, folk, jazz and light classics.

It wasn't all music, though, as there was a significant amount of comedy and sport featured on the schedule as well.

Radio Three succeeded the Third Programme when all the other BBC stations were rebranded in 1967. The station continued to feature cultural fare, classical music, adult educational programmes and sports coverage on Saturday afternoons.

This format remained until 4 April 1970, when there was a structural reorganisation within the BBC. This followed the publication of a controversial policy document entitled *Broadcasting in the Seventies*. As a result, factual content, including documentaries and current affairs, were moved to Radio Four. The report also proposed a large cutback in the number and size of the BBC's orchestras.

The document stated that Radio Three was to have a larger output of serious classical music but with an additional element of cultural speech programmes in the evenings. In reality, this meant that spoken word content was reduced from fourteen hours a week to six.

Nowadays, the station broadcasts regular performances by various orchestras and musicians throughout the year. Productions of both classic plays and newly commissioned dramas remain an integral part of Radio Three's output.

Radio Four continued largely as it did when it was called the Home Service. It remained the BBC's main spoken word channel with a wide variety of drama, documentaries, comedy and current affairs programming. Sport and music are the only subjects that largely fall outside the station's remit but there are exceptions.

One of those exceptions is the long-running music-based *Desert Island Discs*. Each week a guest, usually called a 'castaway' during the programme, is asked to choose eight musical recordings that they would take if they were to be cast away on a desert island and explain the reasons for their choices. At the end of the programme they have to pick the one piece of music they hold in the highest regard. They also have to select a book and luxury item they can take with them. By participating in this process the listener is offered a fascinating insight into the personality and life of the interviewee. More than 3,000 episodes have been transmitted since January 1942. Roy Plomley, who devised it, fronted the programme until his death in 1985. Since then Michael Parkinson,

Sue Lawley, Kirsty Young and Lauren Laverne have each had a spell as presenter of the show.

Although the BBC introduced these wholesale changes to their networks in 1967 because they were forced to, it would be wrong to suggest that they didn't continue to refresh their radio output throughout the 1960s. Sir Hugh Greene became director general in January 1960 and held the post throughout the decade. Greene believed that the corporation should reflect the social changes and attitudes of the sixties.

Greene's appointment coincided with the emergence of a new generation of British writers, journalists and performers who frequently targeted the establishment order. The 'angry young men' were a group of playwrights and novelists who were disillusioned by traditional British society. A new breed of entertainer was also emerging but not from the traditional music hall/variety route. These performers were from Oxford and Cambridge University Revues and used satire as their weapon of choice. Like the 'Angry Young Men', the satirists' target was the establishment. The BBC wholeheartedly encouraged these new movements and was frequently the first to employ the protagonists.

Anarchic radio comedy flourished with new programmes like *Round the Horne* and *I'm Sorry I'll Read That Again*. *Round the Horne* developed from its more traditional predecessor *Beyond Our Ken*. Both series featured the same cast – Hugh Paddick, Betty Marsden, Bill Pertwee, Kenneth Williams and the show's front man, BBC stalwart Kenneth Horne – though *Round the Horne* featured outlandish characters and outrageous scripts full of double entendres.

The cast was given free rein to develop the characters. These included Charles and Fiona, a love-struck couple engaging in stilted, polite dialogues, in scenes that parodied the romantic films of the 1940s. Then there was J. Peasemold Gruntfuttock, a disgusting and degenerate old man, and Rambling Syd Rumpo, an aged folk singer. Probably the most fondly remembered characters were Julian and Sandy, two flamboyantly camp out-of-work actors, with Horne as their unknowing comic foil. The BBC transmitted four series of weekly episodes from 1965 until 1968. A fifth series had been commissioned, but was abandoned after Horne's untimely death in February 1969.

The Origins of BBC Comedy

I'm Sorry I'll Read That Again was an irreverent comedy show, which began in 1967 and ran regularly for seven years. The show relied heavily on the use of puns and included some jokes and catchphrases that would seem politically incorrect by the mid-1990s.

The programme is probably best known as the forerunner of two television comedy successes. The cast included John Cleese, Graeme Garden, Bill Oddie and Tim Brooke-Taylor who had all emerged from student revues. The roots of Monty Python are clearly evident as Graham Chapman and Eric Idle were regular script contributors. The show's creator, Humphrey Barclay, would also go on to create the TV show *Do Not Adjust Your Set*, featuring the rest of the Python team. Graeme Garden, Tim Brooke Taylor and Bill Oddie would go on to create another successful television comedy, *The Goodies*.

I'm Sorry I'll Read That Again spawned a spin-off show, which featured quite a few members of the original cast, at least in the first series. *I'm Sorry I Haven't a Clue* was launched in April 1972 as a parody of radio and TV panel games, and has been broadcast ever since. Described as 'the antidote to panel games', *I'm Sorry I Haven't a Clue* features two teams of two comedians 'given silly things to do' by a chairman. The chairmanship was shared during the first series before Humphrey Lyttleton, the jazz musician, took over permanently. 'Humph', as he was affectionately known, served as chairman until his death in 2008. At first, a trio of hosts serving in tandem replaced Lyttleton: Rob Brydon, Stephen Fry and Jack Dee. Dee took over permanently for the following series and has continued in that role to the present.

Another long-running comedy programme made its debut a few years earlier. *Just a Minute* first aired on Radio Four on 22 December 1967, three months after the station's launch. The object of the game is for panellists to talk for sixty seconds on a particular subject without hesitation, repetition or deviation. The comedy results from attempts to keep within these rules and the witty repartee among the participants.

The game was devised by Ian Messiter as he mused on the top deck of a bus. He recalled an incident from his school days, when a teacher caught him daydreaming in class. As a punishment, the master instructed him to repeat everything he had said in the previous minute without hesitation or repetition. To this, Messiter added a rule prohibiting players from deviating from the given subject, as well as a scoring system based on panellists' challenges.

Nicholas Parsons has appeared on the show since its inception. On nine occasions he has appeared as a panellist but acted as chairman for all the others. There have been five regular contributors in the show's history. Clement Freud and Derek Nimmo appeared from the outset, while Kenneth Williams joined in 1968. Peter Jones made his debut in 1971 and Paul Merton joined in 1988 at the behest of Nicholas Parsons. Merton, a huge fan of the show, replaced Kenneth Williams after his death in 1988. Nimmo died in 1999, Jones in 2000 and Freud in 2009, leaving Merton as the only regular panellist, although he doesn't appear in every edition.

A strong news and current affairs strand had been introduced on the Home Service and many of these continued on Radio Four. *Today*, the network's flagship news programme, began on 28 October 1957. It began as a programme of 'topical talks' to give listeners a morning alternative to light music. *Today* was originally broadcast as two twenty-minute editions slotted in around the existing news bulletins.

In 1963 the programme fell under the auspices of the BBC's Current Affairs department and became more news-orientated. The two editions also became longer, and by the end of the 1960s it had become a single two-hour-long programme that enveloped the news bulletins.

Jack de Manio became its main presenter in 1958. He was held in great affection by listeners, but became infamous for on-air blunders and his inability to tell the time correctly. In 1970 the programme format was changed so that there were two presenters each day. In the late seventies the legendary team of John Timpson and Brian Redhead became established. Since then a succession of presenters such as Libby Purves, John Humphrys, Peter Hobday, James Naughtie, Sue MacGregor and Evan Davis have all fronted this broadcasting institution.

A supplemental nightly news and current affairs programme was added in 1960 and a similar lunchtime show began in 1965.

Shows like *Today, The World at One* and *The World Tonight* remain in their slots today.

Radio One would continue to be the nation's favourite station throughout the 1970s. To fulfil its public service remit, the network was required to broadcast news. However, it was widely thought that the style of traditional BBC news bulletins jarred with the station's format, so *Newsbeat* was introduced on 10 September 1973 in an effort to provide a more tailored news service. DJ Ed 'Stewpot' Stewart was the launch presenter at first but Laurie Mayer and Richard Skinner soon succeeded him.

Although unsubstantiated by the BBC, it is widely thought that the name *Newsbeat* was taken from the Radio Caroline news service of the same name during the 1960s. Roger Gale, a former Radio Caroline North presenter, was one of the show's first producers. The programme has remained a fixture on Radio One ever since.

The BBC was also keen to get back to its roots and, in a move that hearkened back to the early days of 1923, it proposed a network of low-power local stations. In December 1966 the government granted the BBC permission to carry out a two-year experiment in local radio. These stations were to be FM-only and would not be funded by the licence fee. Initially they were financed by local authorities and run by a broadcasting council staffed by local people

The first three stations in Leicester, Sheffield and Liverpool opened in November 1967, followed by five others in Nottingham, Leeds, Brighton, Durham and Stoke-on-Trent. The experiment ended in August 1969 and was deemed to be a success. Despite this, the government was dubious that local authority financial support would be enough to maintain a permanent service and decided that the service would be rolled out using funds from an increased licence fee. Soon another twelve stations were added and today there are over forty in the BBC's local radio network.

To begin with, the local stations only broadcast for a few hours a day and on FM only at low power. In 1972 medium wave transmitters were opened for all of the BBC's local radio stations. In addition, the BBC utilised the international low-power channel, 1484 kilohertz, and frequencies allocated to other countries to plug the gaps.

With an expansion of new BBC services and only a limited amount of frequencies available, it soon became apparent that a reshuffle would be necessary to facilitate this growth. When the Open University started in 1971, it required radio airtime. To reduce disruption to regular programmes, the BBC gave it airtime on the FM outlets of Radios Three and Four, while their regular programmes continued on AM, though most of the allotted airtime on Radio Three was outside regular broadcast hours.

The FM/AM splitting was quickly expanded through the early seventies. Radio Four transferred their schools programmes to FM only in 1973. Meanwhile Radio Three's adult education programmes and Test cricket became available on AM only. Radio Two also split in 1971, with Radio One appropriating its FM frequencies on Saturday afternoons and weekday evenings. From 1973 sports programmes on Radio Two were limited to longwave with FM for its regular programming.

To make way for local radio on medium wave, Radios Three and Four forfeited some of their frequencies on 2 September 1972. Radio Three's relay frequency of 1546 kilohertz was lost, with some transmitters transferring to the main 647 kilohertz frequency. Radio Four's frequencies in England were reduced from six to three. Other frequencies were transferred to local radio and 1088 kilohertz was assigned to the World Service to enable two different programmes to be broadcast to Europe at night.

The remaining Radio Four transmitters were regrouped on 692, 908 and 1052 kilohertz and English regional programmes ended. However, regional news and weather transferred to FM only with regional breakfast programmes for East Anglia and the South West, which didn't have local radio at the time. Radio Four South West was later given a separate medium wave network. Regional opt-outs in Scotland, Wales and Northern Ireland continued as before.

Using fewer frequencies degraded night-time reception, so Radio Four was given a handful of new medium wave relays in 1975. Around that time, Radio Four's Northern Ireland transmitters were taken over by Radio Ulster and Radio Three's Belfast AM relay handed over to Radio Four on a new frequency.

Despite these changes the BBC's radio audience continued to fall throughout the 1970s. Undoubtedly, television was the main cause of this as the population eagerly adopted it as their main source of information and entertainment. Other areas of the media were also finding things tough. Newspapers found it hard to compete with the immediacy of their electronic rival. Cinemas and theatres also saw a sharp decline in attendances during this period.

Listeners wanting to hear an alternative to the BBC's transmissions in the early part of the 1970s had very few choices. There was the non-stop pop and fading signal of Radio Luxembourg or the progressive rock of Radio Geronimo, which had hired airtime from Radio Monte Carlo throughout 1970. In the North Sea, Radio Veronica continued to broadcast to listeners in the Netherlands. The station's success inspired others to attempt a similar service.

The Story of Radio North Sea International

Radio North Sea International, more commonly known by its initials RNI, was to have a troubled and controversial time during its short lifespan. The station's owners were two Swiss businessmen, Erwin Meister and Edwin Bollier, who formed the Mebo Organisation, from the first two letters of their surnames. Meister and Bollier purchased the *Silvretta*, a 630-tonne vessel built in Slikkerveer in the Netherlands, in 1948. The ship was fitted out in Rotterdam to become the most luxurious floating radio station ever.

Renamed the *Mebo II*, the psychedelically painted ship carried a medium wave transmitter, which, at 105 kilowatts, was more than twice the power of the one used by Radio London. In addition, the *Mebo II* also had shortwave and FM transmitters on board and had the capacity to simultaneously broadcast four different stations on four different channels.

Regular programmes in German and English began on 28 February 1970. RNI's English-language programmes were very much aimed at British listeners. The station was always more popular in the United Kingdom than the Netherlands but British listeners were complaining that the station's signal was unsatisfactory. On 23 March 1970 the *Mebo II* sailed to the East coast of England and arrived at its anchorage the following day, six miles from Clacton in international waters.

Several people within the RNI organisation thought that this was a deliberately provocative move and certain to elicit a reaction. Even so, the British authorities' response surprised many. On the evening of 15 April 1970 Harold Wilson's Labour government began jamming the *Mebo II*'s medium wave signal. The Ministry of Posts and Telecommunications, the organisation responsible for policing the UK airwaves, announced that it had received protests from Norway and Italy, the two official users of the frequency.

RNI's next move surprised everybody. Five days before the general election, which was due to be held on 18 June 1970, Radio North Sea International changed its name to the more emotive Radio Caroline International. The station launched a propaganda campaign in support of the Tory party as it was commonly thought that a Conservative administration would look more favourably upon commercial radio.

Many people believe that the Caroline and RNI political campaigns made a difference to the result of the general election. This was an interesting poll: after almost a decade of Labour rule, it was the first one in which eighteen to twenty-one-year-olds could vote. The government had failed to consider that these people had been impressionable teenagers when Caroline was at the peak of its influence, though it's difficult to determine exactly how much the offshore radio anti-Labour campaign affected the result.

Election Day arrived and the polls strongly suggested a Labour win. Nevertheless, against the odds, the Labour party was ousted and the Conservatives won the election. Any hopes, however, that a new administration would result in a change of policy were swiftly dashed and the jamming continued. Two days after the election, Radio Caroline International reverted back to Radio North Sea International.

On 14 July 1970 the Ministry of Posts and Telecommunications issued a statement that the jamming would continue. Then, at 10.55am on 23 July, Radio North Sea International fell silent. The Mebo Organisation sent a telegram to the Ministry of Posts and Telecommunications stating that they had capitulated. Later that day, the *Mebo II*, with the aid of a crane ship, returned to the Dutch coast.

In August RNI was embroiled in another controversial incident. Kees Manders, a nightclub owner who also had links with the Veronica organisation, announced he had become its commercial director. He claimed that he had sold advertising on behalf of RNI and was owed £3,000 in commission. Conversely, RNI's managing director, Larry Tremaine, said nothing had been finalised with Manders. Edwin Bollier and Erwin Meister had initially invited Manders to start a Dutch service from the *Mebo II* and offered him a directorship. For some reason negotiations broke down and Meister and Bollier withdrew the offer.

A few weeks later, on the afternoon of Saturday, 29 August 1970, a salvage tug named *Husky* was spotted approaching the *Mebo II* out of the mist. On board were several hired 'heavies'. Accompanying the tug was a launch, *The Viking*, aboard which were Kees Manders with a woman and a child. In the studio, DJ Andy Archer interrupted his programme to broadcast a series of announcements appealing to listeners to contact RNI's offices and inform them of what appeared to be a menacing situation.

The Viking drew alongside the *Mebo II* and Manders climbed aboard and demanded that the ship be taken into Scheveningen. The captain of the *Mebo II* refused this request. The radio ship was effectively isolated and the crew members were vastly outnumbered by the crewmen on the two boats.

When his demands were refused, Manders returned to the *Viking* issuing a threat to cut the anchor chain and tow the *Mebo II* back to port. Crewmen on board the *Husky* made ready to dowse the radio mast using a water cannon. They were persuaded against this course of action when the DJs warned them over the air that this would send a high voltage current down the water stream and risked killing everyone on board.

Throughout the situation the DJs continued to broadcast an account of what was happening. The crew on board armed themselves with knives and petrol bombs and prepared to repel boarders. Switchboards in London, The Hague, and Zurich received calls from listeners. Station owner Erwin Meister arrived on the scene on board a fast launch, accompanied by the *Euro Trip* tender and other craft. The *Husky* and *Viking* immediately fled the scene. Later that day *van Ness*, a frigate of the Royal Netherlands Navy, stood by to monitor the situation.

Things settled down for a few weeks and RNI was allowed to broadcast uninterrupted. However, at 9pm on 23 September 1970 the tender *De Redder* arrived with a message that the station was to close at 11am the following morning. Programmes continued all night with the DJs on board presenting their final shows.

This shock announcement took the listeners completely by surprise. Presenters Andy Archer and Alan West hosted the final hour. All the other DJs on board at the time said their goodbyes during the show, as did some crew-members. At eleven o'clock the station theme,

Les Reed's *Man of Action*, was played, after which the station's transmitters were silenced. The *Mebo II* was offered up for sale at £800,000. A buyer in Africa was reported to be interested but the deal never went through.

It later transpired that the RNI organisation was deeply in debt. It had agreed to close down for two months in exchange for a payment of one million Dutch Guilders from Radio Veronica. The sum would have to be repaid in full if RNI resumed broadcasting within a certain time frame. To enforce this agreement, a crew from *Veronica* was placed aboard the *Mebo II*, which remained at anchor.

RNI's owners quickly became unhappy with this arrangement and, after two months, attempted to refund the money. The RNI directors took two suitcases stuffed with deutschmarks to the Veronica offices, but Veronica's owners refused, insisting that Meister and Bollier honour the agreement.

On 5 January 1971 the captain was tricked into leaving the ship. In his absence, Edwin Bollier boarded the *Mebo II* and took command of the vessel. As he was the owner this was perfectly legal. In March Veronica sued RNI for breach of contract and a writ was issued claiming that the ship was still on hire to Veronica and that RNI was broadcasting in breach of the agreement. The court eventually found in favour of RNI and the case tarnished Veronica's unblemished reputation.

Things were about to get worse, however. On the evening of Saturday, 15 May 1971, three men left Scheveningen harbour in a rubber dinghy. Three hours later they tied up silently alongside the radio ship. The *Mebo* crew was watching a football match on television so never heard two of the saboteurs climb stealthily aboard. The men placed explosives on a pipeline, which led to an oil tank. They lit a fuse consisting of oil-soaked rags and made a rapid getaway. Minutes later the ship was rocked by a large explosion resulting in a fire that began to spread to the rest of the ship. In the on-air studio, Alan West interrupted his regular English language programme with regular mayday messages.

Captain Hardeveld decided to abandon ship and for the benefit of emergency services, he announced the precise position of the ship over the air. The Swiss engineer also broadcast the same information. At 11.40pm the transmitters were switched off and the crew prepared to abandon ship.

The mayday messages were heard in the Wijsmuller Salvage Control Centre at Ijmuiden and they despatched their tug, *Titan*, to help.

Rival salvage company Smit-Tak also sent their tug, *Smitbank*. The Schevenigen and Noordwijk lifeboats were launched as well. As it turned out RNI's own tender, *Euro Trip*, arrived first and took off ten men, leaving three on board to fight the fire.

The *Euro Trip* and the firefighting tug *Volans* then proceeded to tackle the blaze. By 2.20am the fire was finally extinguished and the crew was allowed to return. The attack didn't achieve its aim as the studios and transmitters were undamaged, allowing the station to continue without interruption. RNI's Dutch service went back on the air the following morning at 6am. Repairs to the vessel, estimated at £28,000, were carried out at sea.

Police launched an immediate investigation and, within a matter of hours, the three attackers had been arrested. On 17 May 1971 Radio Veronica's advertising manager, Norbert Jurgens, was arrested and questioned by Dutch Police. The following day Veronica director Hendrik 'Bull' Verweij was also held. Verweij appeared on Netherlands television to tell how he had paid a man 12,000 Dutch Guilders to force *Mebo II* into territorial waters. Once inside the three-mile limit, the ship would have been liable to arrest or confiscation by creditors.

Five days later the prosecution service formally charged the three men on the dinghy with piracy on the high seas. Norbert Jurgens and Bull Verweij were charged with complicity in the crime and all five men were detained in custody. Support for the five men came from an unlikely source: RNI's owners, Meister and Bollier, didn't want the offenders prosecuted as they thought the resulting publicity would prove disadvantageous to all concerned.

Nevertheless, during September 1971, the five conspirators appeared in court. All five were sentenced to a year in prison. Summing up, Judge Mr van't Veer compared the events that happened to 'gangster methods' and 'totally inadmissible'. The damage to Veronica's reputation was immense.

The bombing of RNI convinced the Dutch authorities that it was time to pass anti-pirate legislation. On 27 May the government announced that it was going to ratify the Strasbourg Convention. Ironically the two bitter rivals, Veronica and RNI, now found themselves working together to try and prevent the bill from becoming law.

Caroline takes to the seas again

Meanwhile, on a grassy quayside in Zaandam Harbour in the North of Holland, a former radio ship had been left neglected and allowed to fall into disrepair. The MV *Mi Amigo* had been one of two Radio Caroline ships impounded by a tender company due to unpaid bills. The other ship, the MV *Caroline*, had been sold for scrap in 1972. The vessel was barely seaworthy but that didn't deter a group of individuals with big plans of their own. It had been sold to the Dutch Free Radio Organisation who planned to turn it into a floating radio museum. The team of volunteers laboured through the summer rectifying what they could of the atrophy and vandalism that the vessel had endured.

However, the pirate radio museum had just been a rather effective cover story. The authorities were told that the 'museum' was being taken to England where it would have a better chance of becoming a profitable tourist attraction. At the beginning of September 1972 the ship quietly left Zaandam and was towed down the canal system towards the North Sea.

The Dutch government had yet to introduce their version of the Marine Offences Act so it was decided to base the operation in the Netherlands. The plan was to introduce a Dutch language service during the day with an English service at night when the signal could reach Britain.

Over the next few years there were numerous test transmissions and various format flips. The slick pop presentation of the 1960s was quickly replaced by progressive rock and eventually a more relaxed freeform album format. Radio Caroline even changed its name to Radio Seagull for a time.

Radio Atlantis operated in 1973 and 1974 from the coast of The Netherlands and Belgium. Belgian businessman Adriaan van Landschoot, who ran the Carnaby chain of companies, owned the station. Van Landschoot wanted to promote his various products, but found there was a lack of radio advertising opportunities on Belgian public radio networks. He also objected that Belgian law prohibited him from establishing a commercial radio station of his own. He decided to organise his own offshore radio station and commenced broadcasting from Radio Caroline's ship on 15 July 1973.

In October 1973 the station's contract with Radio Caroline was terminated, following the collapse of the mast on the MV *Mi Amigo*.

Undaunted, van Landschoot then acquired his own ship, naming it after his wife, the MS *Janine*. The station recommenced broadcasting on 30 December 1973 and transmitted taped Flemish programmes during the day with live English broadcasts at night.

The Dutch Marine Offences Act forced most of the radio stations off the air. Radio Atlantis closed on the evening of 31 August 1974, the day before the act came into force. Radio Veronica and Radio North Sea International also closed, though Radio Caroline and its new Belgian partner, Radio Mi Amigo, defied the act and remained on the air.

The MV *Mi Amigo* moved to an isolated anchorage in the Knock Deep at the mouth of the Thames estuary. This meant that the ship was not visible from either the Essex or Kent coasts. Officially from this point, the boat would ostensibly be operated and supplied from Spain. In practice, the vessel was tendered surreptitiously from ports in Britain, France, Belgium and the Netherlands. Over the next six years the ship endured the ravages of the North Sea. It frequently drifted from its anchorage but somehow managed to recover, but its luck would eventually run out.

On 19 March 1980, during force ten storms, the anchor chain broke and the *Mi Amigo* drifted for ten miles. By the time they succeeded in dropping the emergency anchor, the *Mi Amigo* had run aground on the infamous Long Sand Bank. The coastguard service had been advised of the problem and had arranged for a lifeboat to be placed on standby. The Sheerness lifeboat *Helen Turnbull* arrived just after 6pm and waited about a mile from the *Mi Amigo*.

The Radio Caroline crew discovered that there was water in the bilges and set the pumps running. However, the water continued to force its way into the bottom of the ship and was soon over the bilge plates. For the next few hours the crew continuously pumped water from the leaking vessel. They were convinced that they could still rescue the ship and refused the lifeboat crew's offer of assistance.

Unfortunately, the ship's hull had been eroded wafer-thin due to many years at sea. When a hole appeared in the hull the usual procedure was to stop the inflow of water by whatever means was possible. Usually a wooden cradle was built around the hole and filled with concrete. On this occasion the motion of the ship moving up and down on the sandbank knocked off nearly all of the concrete blocks and the sea started to pour in. After standing by for three hours, the coxswain of the Sheerness lifeboat insisted that the crew leave the ship as parts of it were now waist-deep in water. Reluctantly, the crew decided it was time to leave and informed the coastguard of their decision.

The ship's crew assembled up on the deck and the rescue began in earnest. A first attempt to leave from the stern was found to be impossible. Instead, the lifeboat *Helen Turnbull* approached the *Mi Amigo* on the starboard side as there were rubber tyres suspended there to act as fenders for supply ships. The lifeboat's coxswain asked his men to judge their course and speed so that they would come alongside at the precise moment between the crests of two racing waves. It was a tricky rescue because one error of judgment could lead to the lifeboat being smashed down on the *Mi Amigo*'s deck. The lifeboat crew repeatedly risked their lives in mountainous seas and had to manoeuvre alongside the *Mi Amigo* thirteen times before all the crew could be rescued.

All four members of crew and the ship's canary were successfully lifted off the ship in a rescue operation that lasted just over an hour. The generator was left running to power the pumps, but they couldn't handle the influx of water. The seawater quickly reached the generators, the lights went out, the pumps failed and the *Mi Amigo* sank ten minutes later. Radio Caroline fell silent.

The rise of commercial radio

The BBC had restructured its radio networks to compete with television in the 1960s and these changes had been largely met with approval. Until the early 1970s the BBC had enjoyed a legal monopoly on radio broadcasting in the UK, but its radio audience was about to fragment. Up to this point it had been the policy of both major political parties that radio was to remain under the control of the BBC. Its only rivals had been sporadic broadcasts from European and offshore stations, but now there was a serious competitor on the horizon.

A change of government occurred in 1970, which saw the passing of Harold Wilson's Labour administration to Edward Heath's Conservative government. This new administration looked upon the introduction of commercial radio much more favourably. The new Minister of Post and Telecommunications, Christopher Chataway, announced a bill to allow for the introduction of commercial radio in the United Kingdom.

This service would be planned and regulated in a similar manner to the existing ITV service and would compete directly with the recently developed BBC Local Radio services.

The Sound Broadcasting Act was passed on 12 July 1972 and the Independent Television Authority (ITA) was renamed the Independent Broadcasting Authority (IBA) on that day. The authority immediately began to plan the new service and placed advertisements encouraging interested groups to apply for medium-term contracts to provide programmes in given areas.

The first major areas to be advertised were London and Glasgow. There were two franchises available in London: news and information and a general entertainment service. The London news franchise was awarded to London Broadcasting Company (LBC) and they began broadcasting on 8 October 1973. Capital Radio was awarded the general entertainment contract and their first broadcast took place eight days later. The Glasgow contract was won by Radio Clyde who made their debut on New Year's Eve 1973.

Altogether, nineteen contracts were awarded between 1973 and 1976. The development of ILR paused at this point as the Labour Party had been returned to power. The Labour government was wary of commercial organisations running radio stations and so temporarily halted any further development of the Independent Local Radio system.

Surprisingly, a part of the British Isles had already experienced commercial radio nearly a decade earlier. In 1959 the Tynwald, the government of The Isle Of Man, passed a bill to establish an independent commercially funded radio station on the island. They thought that a radio station would greatly benefit the community and the economy. Although the Isle of Man is self-governing, the Tynwald was required to apply to the UK authorities for a transmitting licence. The British government opposed the project but eventually the necessary permissions were granted.

The station went on air using an FM frequency of 89.0 megahertz in June 1964. The first broadcast was a commentary about the Isle of Man TT race. They transmitted from studios in a caravan just outside Douglas with a temporary aerial mast located next to the caravan. In October a medium wave transmitter was established on 1595 kilohertz using a mast at Foxdale. The temporary FM mast was soon replaced with a permanent installation on Snaefell. A relay station on 91.2 megahertz was added to improve reception in Douglas in 1969.

In 1965 the station moved to permanent studios on the Douglas seafront, and a second medium wavelength of 1292 kilohertz was allocated to improve the coverage, which had been limited by the high frequency of the 1594 kilohertz service. However, the 1292 kilohertz service could only be used during daylight hours, so listeners had to retune to 1594 kilohertz when darkness fell. The programmes from Manx Radio were not only popular on the island but also with listeners in northern Britain.

Meanwhile, back at Broadcasting House, the BBC had managed to stave off competition from television but the introduction of new commercial radio stations in the 1970s had seriously affected the corporation's audience share. A radical shake-up of frequencies was on the way and the management would use this as an opportunity to rearrange their radio furniture.

Chapter 10

All Change

On 23 November 1978 there was a comprehensive reorganisation of medium and longwave frequencies. These changes were intended to make more efficient use of frequencies and to reduce interference. All medium wave stations were realigned into nine kilohertz spacing. For local BBC and ILR stations this was just a case of altering frequency by one or two kilohertz: for example, Capital Radio moved from 1546 to 1548 kilohertz, while LBC and Clyde moved from 1151 to 1152 kilohertz. One exception was Manx Radio on the Isle of Man, which had its day and night-time frequencies of 1295 and 1594 kilohertz replaced by one frequency of 1368 kilohertz, which would be used round the clock.

The big changes in the UK were for BBC national radio. The BBC World Service and Radio Four vacated their frequencies, leaving Radio One free to transfer to 275 and 285m (1089 and 1053 kilohertz). The pop music network also benefited from a network of much more powerful transmitters.

Coverage of Radios One and Four was much improved by the changes, though Radio Four listeners without longwave had to replace their radios. Radio Two's reception during the day was enhanced in some parts of the country, with only a few places losing out. However, night-time reception was diminished.

Only Radio Three suffered a major drop in reception quality. It lost the 647 kilohertz signal from Daventry and moved to Radio One's old slot on 1215 kilohertz, which suffered co-channel interference in many parts of the country. The new 648 kilohertz channel was reallocated to the BBC World Service, and was transmitted to Europe from the Orfordness transmitting station. There were some positives, however. It had the smallest audience but the vast majority used FM anyway. The main service was now transmitted on AM and FM full-time. The early evening adult education service on medium wave was transferred to Radio Four FM at the weekends. This meant the only AM opt-out was the Test cricket coverage.

Radio Four left medium wave and transferred to longwave 198 kilohertz. Radio Two in turn left 200 kilohertz longwave and moved to two of the old Radio Four wavelengths: 693 and 909 kilohertz. This move paved the way for Radio Two to broadcast round the clock in January 1979.

Radio Four's move to being a fully national service on longwave meant that spare frequencies were available. The BBC seized the opportunity to introduce Radio Scotland and Radio Wales as totally separate networks. Up to this point, listeners in Scotland and Wales heard only opt-out programming from Radio Four.

People desiring an alternative to the BBC and commercial stations were treated to another golden offshore period in the mid-eighties. When the MV *Mi Amigo* sank at the beginning of the decade many people thought that this was the end for Radio Caroline. They were wrong. The former trawler and salvage vessel, MV *Ross Revenge*, was built in Bremerhaven, Germany, in 1960. After a twenty-year career at sea, the ship was sold to a mysterious buyer for an undisclosed sum. She was taken to Santander in Spain, where work started to convert her into the new Radio Caroline ship. Then, on 20 August 1983, with an enormous 90-m transmission mast, the *Ross Revenge* started broadcasting to a large pan-European audience.

Another station, Laser 558, joined Caroline in the Knock Deep channel in international waters. This station was launched in May 1984 by a consortium of British and American business and broadcasting executives from the radio ship, *The Communicator*. Laser 558 used presenters recruited from the USA. Within months the station had gained a sizeable audience, popular because of its non-stop music format.

This revival of offshore radio prompted the Department of Trade and Industry to instigate a blockade designed to starve the pirate stations out of existence. They chartered the ocean-going launch, *Dioptric Surveyor*, and moored it in the Knock Deep to record and monitor activities around the radio ships.

Radio Caroline chose a low-key approach and rarely mentioned their new neighbours on air. Laser 558 chose an alternative action. They regularly reported on the movements of the *Dioptric Surveyor* during their programmes. Disc jockey Charlie Wolf, in particular, regularly ridiculed activities taking place on board the surveillance ship. He nicknamed the vessel 'Moronic Surveyor' and also gave the whole operation the title 'Eurosiege 85'. Laser 558 also produced a number of parody commercials for 'Anoraks DTI' satirising the DTI surveillance activities and the cost of the operation to the taxpayer.

The blockade swiftly became highly effective. After only ten days, six vessels had been reported for allegedly supplying the two radio ships. Four had been reported to British police for possible offences under the Marine etc. Broadcasting (Offences) Act, another from Holland had been reported to Dutch police and the sixth had been reported to the Director of Public Prosecutions. Laser 558 reduced its programme hours at the beginning of September 1985 due to continuing staff and supply shortages.

An indication that 'Euroseige 85' was set to continue throughout the rough winter weather in the North Sea came on 1 November 1985 when the *Dioptric Surveyor* was replaced by a larger surveillance vessel, the *Gardline Tracker*. Ironically, this vessel was a sister ship to Laser's vessel. *The Communicator,* which had previously been known as *Gardline Seeker*.

By late autumn 1985 it was reported that Laser 558 was in serious financial difficulties. A lack of advertising starved the station off the air in November 1985. In 1986 an attempt was made to return as Laser Hot Hits, but the same problems quickly arose.

The DTI's blockade was called off and Radio Caroline continued to broadcast from international waters for the rest of the decade. Her final pirate broadcast took place in November 1990. The *Ross Revenge* ran aground on the Goodwin Sands in November 1991, bringing the era of offshore pirate radio in Europe to an end. The ship was, however, salvaged and is now preserved by a group of supporters and enthusiasts called the Caroline Support Group.

There was another pirate boom in the Republic of Ireland. In 1972 RTE had introduced the Irish language service, Radio na Gaeltachta. The popular music station RTE Radio Two, later renamed 2FM, followed seven years later. However, there were no commercial stations, and dissatisfaction with the state broadcaster had led to a profusion of unlicensed pirates starting up all over the country. During the 1980s, Pirate stations like Radio Nova and Sunshine Radio were extremely popular and dominated the Irish airwaves. To this day, Sunshine Radio still holds the highest ratings of any Dublin radio station.

At the beginning of the 1980s, FM was prevalent on portable radios and remained an option for cars, although it didn't become the norm for vehicles until the end of that decade. FM hadn't really caught on in the UK and lagged well behind other countries There were two major problems that explain this: poor reception and lack of spectrum.

The FM radio system was originally designed for reception of mono signals via roof aerials, so horizontal polarisation was chosen to give better directional reception. Unfortunately, the upright aerials used for car and portable radios are more suited to horizontal polarisation as it gives stronger signal strength near the ground.

At this point, all independent, and some BBC local radio stations had converted to mixed horizontal and vertical polarisation, which gave much better reception. Through the 1980s the BBC doubled the power of its FM transmitters and switched to mixed polarisation. The addition of numerous low- and medium-power relay transmitters also improved the situation. Nearly all areas now receive good mono reception on portables and in cars, though stereo reception can still be patchy without a roof aerial.

Throughout the decade it was decided to extend the FM band in the UK. This was done in stages, with the first chunk of spectrum allocated for use in 1983. This was primarily used to introduce new commercial stations, but the BBC also utilised the extra capacity. The organisation expanded some of its city stations into county-wide stations, with name changes where appropriate.

During 1986 and the first half of 1987 more than half of the local radio transmitters changed frequency as both the upper and lower local bands were divided into separate commercial and BBC sections. This shake-up enabled most of the older commercial stations to increase their transmitter powers. This frequency reallocation also allowed the whole FM band in Ireland, both the North and the Republic, to expand. Prior to this, there had only been enough space for two national networks in the country.

The next piece of FM spectrum was not due to be released for broadcasting until the beginning of 1990, although a few transmitters had started up before this. This section was given to the BBC for Radio One and a new national commercial station. The top end of the FM band was finally allocated to broadcasting in 1995. The original intention for this part of the spectrum was to offer extra transmitters for the five national networks. However, in the end it was allotted to regional and local commercial radio instead. The lower part of the FM band was assigned to temporary low-power stations. By the beginning of the 1990s, FM had become the foremost medium for radio listening in the United Kingdom.

In 1979 the Conservative government, led by Margaret Thatcher, swept to power and the expansion of the Independent Local Radio network resumed. A second block of ILR franchises was issued between 1980 and 1984. Radio Mercury was the last of the original stations to go on air. Their opening broadcast was transmitted on 20 October 1984.

Despite this ruling, there were a few casualties along the way. Centre Radio went into receivership on 6 October 1983, and in Wales, CBC combined with Gwent Broadcasting to become Red Dragon Radio. This entity became a much more successful venture. In 1985 Radio West in Bristol merged with Wiltshire Radio and was renamed GWR. A decade later GWR would introduce network programming and revolutionise British local radio forever.

Nevertheless, despite these setbacks, the expansion of local radio didn't stop there. In 1986 the IBA sanctioned the idea that different services could be broadcast on each station's FM and AM frequency. To test the situation, the IBA and the Home Office created a two-year experiment. Selected stations were asked to produce different programming simultaneously on the AM and FM frequencies.

Marcher Sound provided separate programming in Welsh, and the Radio Trent Group vastly expanded its Asian programming. Meanwhile, Viking Radio offered their listeners a choice of rugby league or country music. Piccadilly Radio broadcast the Halle Proms live, and Capital Radio trialled CFM, a special service featuring adult-oriented rock.

These experimental split-frequency broadcasts were declared a big success. In 1988 the government effectively declared the end of simulcasting. They issued a decree that meant ILR stations had to provide different programming on each of their wavebands. If they failed to do this they would forfeit permission to broadcast on both FM and AM.

The first station to permanently split their frequencies was Guildford's County Sound, which rebranded the FM output as Premier Radio and turned the AM output into a new golden oldies station, County Sound Gold, in 1988. Other stations swiftly followed suit.

In addition to the ever-expanding list of official stations, there continued to be some illegal alternatives. Despite the efforts of the authorities, pirates were alive and well with several stations regularly on the air around the country. Unlike their nautical predecessors, however, these unlicensed operations were firmly land-based. Furthermore, the individuals behind these stations were not businessmen hoping to make money. Instead, early stations such as Radio Free London, Radio Jackie and Kaleidoscope were run by disgruntled fans of offshore radio who were unhappy with the official alternatives.

These early land-based stations proliferated in several major conurbations and broadcast mainly on medium wave. All of the programming was pre-recorded and transmitted from a remote area. They would string up a wire aerial between tall objects such as trees or lamp posts, then attach a cassette player to a home-made transmitter powered by a car battery. This primitive set-up was highly effective and could generate a good signal that propagated over a wide area.

These land-based pirates established a strong foothold in the London area during the late sixties and early seventies, and got more sophisticated as time went on. Radio Jackie was really a 'community' radio station and campaigned vigorously for a licence to broadcast in their native South-west London. They did have a year or so of actual live, round-the-clock broadcasting but eventually closed down after a particularly heavy raid by the Home Office in 1985.

Eventually the pirates began to move away from being offshore tribute stations. Instead they began to target a niche audience of music fans who felt ignored by mainstream radio. Broadcasting mainly from tower blocks, these onshore pirates were the pioneers of the pirate scene that exists today. These stations thrived during the eighties, playing mainly reggae and soul music. At their height there were more than fifty stations broadcasting to London including Invicta 94.2, Horizon, LWR and Solar.

There was also an abundance of unlicensed stations in Ireland. The Radio and Television Act (1988) was introduced and the Irish pirates were silenced overnight. From 1989, licenses were awarded on a franchise system and for the first time new, legal stations, not owned by RTE, started up.

The British government also introduced tough new legislation in December 1988, and a lot of the original land-based pirates closed down.

Yet a third generation of pirates such as Sunrise, Centreforce, Fantasy 98.1 and Dance FM gradually replaced them. This coincided with the emergence of dance music as a major influence in the UK. With the rise of house, hardcore and, later, drum & bass, a fourth generation of pirates like Dream, Kool FM, Rush and Pulse FM took control of the London airwaves. BBC Radio One had also recognised the importance of the dance scene and recruited many former pirate DJs to front their specialist output.

A couple of pirates have successfully made the transition to fully licensed stations. During its unofficial period, Kiss FM claimed a massive half a million listeners with its mix of soul, house and hip-hop. The station closed down in 1988 with the goal of obtaining an official licence. Kiss FM eventually succeeded and was granted a licence in September 1990 on its second attempt.

Radio Jackie, the early pirate pioneer, applied for a regional licence when one became available for their area in 1996. However, the new FM licence was awarded to Thames Radio instead. After a few years, Thames Radio experienced financial problems and the station was put up for sale. The original Jackie management team swooped to purchase the loss-making station, which was relaunched as Radio Jackie on 19 October 2003.

The implementation of the 1990 Broadcasting Act led to several major radio developments for the BBC and commercial radio sectors. The government launched a radio spectrum audit and decreed that the BBC would have to end simulcasting its services on both AM and FM frequencies. This meant that the BBC would relinquish some AM frequencies and Radio One and Radio Two would broadcast on FM only. Although the execution of this policy meant that a number of programmes, which were previously broadcast as opt-outs on one frequency only, would otherwise have been left without a home.

The BBC then introduced Radio Five, which began at 9am on 27 August 1990. The new network broadcast on the old Radio Two AM frequencies of 693 and 909 kilohertz. The first voice heard on the station belonged to a five-year-old boy called Andrew Kelly who uttered the words. 'Good morning and welcome to Radio Five.' There followed a pre-recorded programme called *Take Five* introduced by Bruno Brookes. Many local broadcasters received their first national exposure on the network. Local radio stalwarts like Danny Baker, Mark Radcliffe and Martin Kelner all made their national debut on Radio Five.

Broadcasting for around eighteen hours per day, the new network was to carry a variety of sports, children's, educational and minority interest programmes. However, Radio Five was not a ratings success due to its very uneven programme mix. Many thought the station broadcast programming the other four main BBC stations didn't want. Even the BBC's director general at the time criticised it openly. John Birt said that the station sounded 'improvised and disjointed'.

In January 1991, Operation Desert Storm was launched in response to the Iraqi invasion of Kuwait. Radio Four's FM frequencies were used to provide an all-news network for the coverage of the war. 'Radio Four News FM' was well received and the positive response to the rolling news format prompted the BBC to look into the possibility of providing a full-time news station. It was decided that the station would relaunch as a combined news and sport channel. The 'old' Radio Five signed off at midnight on Sunday, 27 March 1994, and the new Radio Five Live began its twenty-four-hour service at 5am the following day.

Perhaps the biggest development to come from the 1990 Broadcasting Act was the abolition of the IBA and the introduction of a new regulator. The Radio Authority had a different remit to its predecessor. It would be allowed to issue licences to the highest bidder and promote the development of commercial radio choice. This led to the awarding of three national contracts for Independent National Radio.

INR1 was the only one of the three new franchises that would broadcast on FM. This franchise was advertised as a non-pop licence, and was awarded to Classic FM, which launched on 7 September 1992. Classic FM enjoyed almost instantaneous success, providing listeners with a quality programme of 'accessible' classical and orchestral music, and a comprehensive news service.

The other two franchises would not fare so well. INR2 was allocated the former medium wave frequency of BBC Radio Three. This licence was awarded to Virgin 1215, with a service of rock-orientated music. The new service began on 30 April 1993 and was popular among rock fans. It was not, however, the financial success a national music station could have been and the adventurous music policy was increasingly diluted and became more conventional. The station has since become Absolute Radio and has utilised new broadcasting technologies to launch several complementary digital services.

The Radio Authority awarded the third franchise to Talk Radio UK, which started transmissions on Valentine's Day in 1995. The station utilised the old Radio One AM outlets vacated by the BBC's pop network in 1994. At its launch, the station employed many 'shock jocks' whose aim was to provoke debate by being deliberately outrageous. However, this American style of presentation was unpopular with British audiences and prompted the station to alter the format fairly swiftly. The confrontational approach of presentation was replaced by more traditional 'phone-in' shows. Talk Radio UK employed experienced speech presenters such as James Whale, Paul Ross, Mike Dickin and Nick Abbot.

Despite gaining a fairly sizeable audience, Talk Radio seemed unable to generate a profit. In 1999 the station was taken over by The Wireless Group, in partnership with media mogul Rupert Murdoch. The station relaunched as 'TalkSport', with former newspaper editor Kelvin Mackenzie at the helm. Under his leadership the station enjoyed a renaissance; audiences increased considerably and this led to the station becoming profitable. In 2005 The Wireless Group was sold to Ulster Television and TalkSport became part of UTV Radio.

The early 1990s were tough for the BBC, whose position as a public service broadcaster was under constant scrutiny. Many critics focused their ire on Radio One, believing it was senseless for the BBC to provide a 'throwaway' popular music station. Even some government ministers suggested the corporation should privatise the network.

Although originally launched as a youth station, by the early 1990s Radio One's loyal listeners, and most of its presenters, had aged with the station over its long history. Therefore, when Matthew Bannister took over as programme controller in October 1993, his aim was to reposition the station's core demographic and make it appeal to a younger age group.

Many long-serving 'personality' presenters, such as Dave Lee Travis, Simon Bates, Alan Freeman and Johnnie Walker either left the station or were sacked. In their place Bannister parachuted in new, younger DJs who would appeal to his target audience and make the station relevant again. Then, in January 1995, older music (usually anything recorded before 1990) was omitted from the daytime playlist and a greater emphasis was placed on promoting new music from up and coming artists.

There is no doubt that Matthew Bannister needed to act to make Radio One relevant to its target audience. However, a gradual evolution rather than a quick revolution may have had better results. Johnny Beerling, who was Bannister's predecessor, preferred a more cautious approach and made changes gradually. By implementing these changes rapidly, Bannister sacrificed a sizeable part of the listenership. Over the next few years Radio One lost around five million listeners, many of whom felt alienated as the new music format left them with nowhere to go.

At that time, Radio Two was sticking doggedly to a format that appealed mainly to people who had been listening since the days of the Light Programme. This left the way clear for commercial radio to clean up at the expense of Radio One. They deliberately targeted the disenfranchised audience and enjoyed a massive increase in audience share.

In 1996, after several years in the doldrums, Radio Two was about to reposition itself. James Moir was appointed as the station's controller and he set about building an audience. He freshened up the daytime playlist by adopting a more contemporary feel. In addition, he added several well-known presenters and popular television personalities to the mix. Established names such as Jonathan Ross, Mark Radcliffe, Bob Harris and Steve Wright joined existing Radio Two personalities such as Terry Wogan and Ken Bruce on the schedule.

Unlike Matthew Bannister, his Radio One counterpart, Moir progressed stealthily by making small changes over a prolonged period. This strategy managed to retain the existing audience while also attracting a much larger one. Radio Two went from a station languishing in a broadcasting backwater to the most popular one in the country. The station's appeal remains broad and deep, with accessible daytime programmes and specialist programmes at night.

The Radio Authority began to licence low-power temporary radio stations for special events, operating for up to twenty-eight days a year. They also reduced the criteria for a 'viable service area' with the introduction of small-scale local licences for towns, villages and special interest groups.

At this point in time the AM waveband had become unpopular with radio groups and the majority of new stations were awarded an FM licence only, even when an AM licence was jointly available. The Radio Authority also introduced regional stations and began to licence the commercial Digital Audio Broadcasting (DAB) multiplexes in October 1998.

The Radio Authority's lighter touch also meant a relaxation of ownership rules. Tentatively, stations began to form small local groups to take advantage of the financial opportunities this produced. Stations in the group could share news and sales operations. Gradually, several distinct large radio groups evolved. These included EMAP (Now Bauer) and The Wireless Group (later UTV). Capital Radio merged with the GWR group in 2005 to form Gcap Media plc (now owned by Global Radio).

As these acquisitions and mergers gathered pace, specialist local programmes were dropped and output began being shared around the networks. Most of the localised medium wave 'gold' or 'classic hits' stations disappeared and became either 'Capital Gold', 'Classic Gold' or 'Magic'. The networking of programmes would proliferate rapidly and would not just be confined to medium wave outlets.

Even with the best of intentions, existing radio services can't hope to cater for the full range of listeners' interests. Special interest material will always be marginalised in a mainstream schedule. The homogenised output of the radio networks has left many listeners disenfranchised and this led to many groups calling for a third tier of radio broadcasting to be introduced.

There were two distinct factions promoting this third tier. Firstly, there were those who desired a simple deregulation of the airwaves. When this happened in France and Italy it led to even more similar-sounding stations being introduced and a lack of real choice for the listener. The authorities have largely ignored the groups seeking deregulation, preferring the second faction instead: individuals or ethnic groups seeking to introduce community radio to their area.

The meaning of community radio is straightforward: It is radio that is owned, managed and made by its audience. Any member of the public can become a member of the group running their local station and make and broadcast programmes, without being filtered through the mediating hands of the professionals.

Community radio stations were in operation on cable systems as far back as 1978. In the late eighties the then-newly formed Radio Authority awarded licences (termed 'Incremental' by the outgoing Independent Broadcasting Authority) to a number of new, ex-pirate and cable-based community ventures. These stations were introduced to provide extra specialist stations for areas with an existing commercial radio station.

The first four incremental stations were granted licences in 1989: they were Sunset 102 in Manchester, CentreSound in Stirling, FTP in Bristol and Sunrise Radio in West London. They were forbidden to seek public funding and advertising was their only source of income. Not surprisingly, many of these new stations crashed and burned very quickly. FTP and CentreSound failed within a year of launching and, after a managerial restructure, were relaunched as Galaxy 97.2 and Central FM respectively. Sunset Radio lasted until October 1993 before filing for bankruptcy. Faze FM was awarded the re-advertised 102.0 FM licence and operated under the name Kiss 102. Rather than a community station, Kiss was a dance music station and was later sold to Galaxy Radio.

The only one of the first four to survive intact was Sunrise Radio. Sunrise quickly expanded, first into Bradford, taking over the licence for the failed Bradford Community Radio. A satellite service followed in 1991. Then in 1994, Sunrise Radio won the Radio Authority licensing process to expand its programming across the whole of London and the surrounding counties from the 50-kilowatt transmitter at Brookman's Park.

Some other stations have survived and thrived. Spectrum Radio, a multi-ethnic foreign language station, launched originally on 558 kilohertz and is still on air, albeit only on DAB. Another London station served a specialist interest group rather than a minority ethnic audience. London Jazz Radio launched as Jazz FM with a professional team of presenters and a signal similar to Capital Radio.

For a while things seemed to be going well, the organisation also won the North West of England regional licence. However, Jazz FM could never make enough money to be sustainable in that form. The station was eventually purchased by GMG Media, who changed the format to oldies and tuneful pop and renamed it Smooth Radio. Jazz FM was relaunched in 2008 as an online and digital station.

Many other incremental stations were unsuccessful. RWL (Radio West Lothian) 1368, which was based in Bathgate, failed very quickly in 1990. East End Radio in Glasgow lost its licence several years after going on air. Radio Harmony, the ethnic station in Coventry, remained on air for several years before being taken over by KIX-96, a mainstream pop music station. It was a similar story with Belfast Community Radio (BCR), which, after six years of struggling, flipped to a mainstream pop format in 1996.

One long-lasting and beloved radio station would not make it to the twenty-first century. Radio Luxembourg had battled on bravely but its weak signal and the proliferation of new competition had meant that the station's listenership had plummeted. Stephen Williams, who had been the station's first announcer, returned one last time on 30 December 1991, when Radio Luxembourg closed down its medium wave English service on 208m. The eighty-three-year-old veteran made the final announcement, 'Good luck, good listening and goodbye.'

Radio had managed to retain a sizeable audience during the last half of the twentieth century despite the rising dominance of television. It did this by constantly evolving and adapting to audience needs and economic trends. Yet, as the new millennium dawned, radio would face stiff competition from more emerging technologies.

Chapter 11

Radio in the Twenty-First Century

At the beginning of the twenty-first century there was a radical shift in how listeners received music and information. Young people in particular saw radio as outmoded. They preferred to listen to music or access news and information via portable music players and the Internet.

There has been a noticeable drift away from AM to FM broadcasts since the new century began. Audiences for long, medium and shortwave broadcasts continue to fall. In fact, many international broadcasters have ceased their shortwave transmissions and increasingly the only things that can be heard on this band are radio hams, utility and time stations.

However, this doesn't mean that radio has finally been eclipsed. Radio simply did what it did when confronted by the emergence of television in the 1950s: adapt to survive. Radio moved from being a linear broadcast on one device to unique audio content delivered via multiple platforms. Emerging technologies such as satellite, DAB and the Internet offered radio stations a multi-platform future. As digital forms of radio proliferate, listeners will enjoy an abundance of new programming.

Radio has fought off competition from broadband Internet, YouTube, Spotify and Facebook. Indeed, its strength is that you can still enjoy radio while you do many of these things. Podcasts or 'listen on demand' facilities on radio station websites are enabling audiences who didn't catch the original broadcast to listen later at a time that suits them.

Most traditional radio stations now have an Internet outlet. Internet radio utilises streaming media, presenting listeners with a continuous stream of audio that cannot be paused or replayed. Streaming on the Internet is usually referred to as webcasting since it's not transmitted using traditional broadcast methods. Internet Radio is distinct from on-demand file serving and podcasting, which involves downloading the material rather than streaming.

It's uncertain who first thought of Internet radio, but records show that the first actual Internet radio station was by a non-profit organisation called The Internet Multi-casting Company of Washington, which began in 1993. They helped Carl Malmud launch an online chat show called *Internet Talk Radio*, which featured interviews with notable people in the field of science and technology. However, this couldn't really be classed as Internet radio as, conceptually, it was just a radio show on the Internet.

There were several Internet-based firsts the following year. In March 1994 an unofficial automated rebroadcast of Irish radio news was setup as the 'RTE to Everywhere Project', which allowed Irish people across the world to access daily radio news from home. In November 1994 a Rolling Stones concert was hailed as the 'first major cyberspace multicast concert', though a band called Severe Tire Damage had already streamed one of their shows five months earlier.

On 7 November 1994 WXYC became the first traditional radio station to commence broadcasting on the Internet. The college radio station, based at the University of North Carolina, became an Internet pioneer when Michael Shoffner set up their streaming service. He had begun testing as early as August 1994 using an FM radio connected to an Internet video conferencing site.

They were joined a week later by WREK in Atlanta, which was staffed by students at the Georgia Institute of Technology. These stations didn't stream full time but KJHK, located at the University of Kansas, became the first station to stream live and continuously over the Internet on 3 December 1994.

All these stations used their own streaming software but Progressive Networks released RealAudio in 1995. This software took advantage of the latest advances in digital compression and delivered AM radio-quality sound in real time. Eventually, companies such as Nullsoft and Microsoft released their own streaming audio players and many web-based radio stations began to proliferate.

In 1995 Scott Bourne founded NetRadio.com as the world's first Internet-only radio network, and the first to be officially licensed by the performance-rights organisation ASCAP. NetRadio.com was a true digital pioneer. Most of the current Internet radio providers used the company's business model as a template on which to build their own success.

In March 1996 Virgin Radio became the first European radio station to broadcast its full schedule live on the Internet. It broadcast its FM signal live from the source and simultaneously on the Internet twenty-four hours a day. Nowadays, even small community and hospital stations have an Internet stream and can subsequently be heard well outside of their locality.

As DSL and broadband Internet replaced the old dial-up phone line connections, Internet radio became more popular. Many people believe that the Internet will become the main delivery method for radio in the future. As data costs continue to fall, listening to a radio station via an Internet stream will indeed become more popular and widespread. However, many see Internet streaming as just one part of a multi-platform future for radio. It's far too early to determine which method will become the dominant one.

In 2004 radio in Britain fell under the control of the Office of Communications, commonly known as Ofcom. This new regulator also replaced the Independent Television Commission, the Broadcasting Standards Council, the Radio Communications Agency and the Office of Telecommunications.

In addition to AM and FM transmission methods, radio stations gained a digital alternative. The Broadcasting Act of 1996 allowed the introduction of national, regional and local digital broadcasting to Britain. Traditional broadcasting methods used up a comparatively large amount of spectrum for a relatively small number of stations. Digital audio broadcasting (DAB) combines multiple audio streams onto a relatively narrow band centred on a single broadcast frequency called a DAB ensemble or multiplex.

The BBC had started digital radio test transmissions from Crystal Palace as far back as 1990, and permanent transmissions to London began in September 1995. With the expansion of its network in the spring of 1998, the BBC's DAB multiplex was available to sixty-five per cent of the UK population by 2001, and to eighty-five per cent by 2004. All the BBC's existing radio services appeared on the ensemble, including the World Service. This was the first time the station was available to a domestic audience in crystal-clear quality. Though there were more digital delights to come.

The arrival of five new services in 2002 marked the largest expansion of radio in the BBC's history. 5 Live Sports Extra, a companion station to 5 Live, was the first to launch in February of that year. The part-time station provided uninterrupted sports coverage when there were clashes of major events where the BBC had the broadcasting rights.

BBC 6 Music, the BBC's first new national music station for thirty-two years, followed in March 2002. This station filled the gap between Radios One and Two, offering an alternative mix of classic and contemporary rock music. In addition, other specialist genres such as jazz, funk, blues and folk would also be featured. The BBC's vast archives of live music sessions would form an important part of BBC 6 Music's programmes.

In February 2010 it was announced that the BBC were thinking of closing the station as part of a bid to cut expenditure. Immediately, a high-profile campaign was launched to oppose closure of the station. Five months later, the BBC Trust announced that it was not convinced by the proposal and that the station would stay. In April 2011 the network was to be rebranded to BBC Radio 6 Music.

1Xtra, a sister station to Radio One, was launched on 16 August 2002. This station was aimed at young fans of cutting-edge urban music, such as rap, hip-hop and R&B. As well as contemporary black music, the station featured a dedicated news service and regular speech-based programmes.

The BBC Asian Network, already a successful regional service on FM in the West Midlands and the North, launched nationwide on 28 October 2002. Offering a mix of news and music, this station was aimed at the diverse Asian communities across the UK. The station broadcasts mainly in English, but also features programmes in various Asian languages.

BBC 7, a new speech station, was launched in December 2002. This station was the principal broadcasting outlet for the BBC's archive of spoken-word entertainment and many old favourites like *Hancock's Half Hour* and *Dick Barton* were rebroadcast to a modern audience. The station was rebranded as BBC Radio Four Extra on 2 April 2011 to bring the station closer to its sister station, Radio Four.

BBC Radio Five Live also gained a sister station that specialised in extended coverage of additional sports. BBC Five Live Sports Extra began broadcasting at 2.30pm on 2 February 2002. Juliette Ferrington introduced the first programme, which was commentary of a football match between Manchester United and Sunderland.

The new station became the digital home of *Test Match Special*, which had been a mainstay of the BBC since it started in 1957.

The programme started on the AM frequencies of the Third Programme before moving to FM in the summer of 1992 when Radio Three lost its AM frequency. The following summer the morning play was on Radio 5, before transferring to Radio 3 for the afternoon session. At the inception of Radio 5 Live the programme moved to its present home on Radio 4 longwave.

The first national licence for DAB from the Radio Authority was advertised in 1998 and only one applicant applied. The licence was awarded to the GWR Group and NTL Broadcast. The two companies formed the Digital One multiplex, which began broadcasting on 15 November 1999. The ensemble has grown and is currently available to over ninety per cent of the UK population. The United Kingdom presently has the world's biggest digital radio network, with 103 transmitters, two national multiplexes and forty-eight local and regional ensembles broadcasting over 250 commercial and thirty-four BBC radio stations.

Despite the existence of the DAB network, the medium initially struggled to gain mass acceptance by consumers. Cost was a major issue in the early days. Digital radios were first sold as car radios in 1997, priced around £800. Hi-fi tuners costing up to £2,000 were released two years later. In 2002, Pure Digital's Evoke series of radios broke the £100 price barrier and the cost of a DAB radio has since fallen to around £40.

Lower prices, new radio stations, and marketing have increased the uptake of DAB radio in the UK. DAB was promoted in Britain as having two major advantages over analogue radio broadcasting. In using compression technology, parts of the audio spectrum that cannot be heard by humans are discarded, meaning less data needs to be sent over the air. This, as well as multiplexing technology, allows a number of channels to be broadcast together on one frequency as opposed to one channel for analogue radio broadcasts. This meant that broadcasters were able to launch exclusive digital radio stations alongside their existing analogue stations. Broadcasters also state that DAB offers better reception and is resistant to the interference which other broadcast media are susceptible to. DAB radios also come with additional features such as scrolling text, providing information such as breaking news, travel information or the latest track information.

DAB's critics say that the audio quality on DAB is lower than on FM. Also, a large and growing number of music stations are transmitting in mono. Indeed, the bit rates used by the radio stations on other digital platforms, such as cable, terrestrial and satellite, are usually higher than on DAB, so the audio quality is also higher. However, an Ofcom survey, undertaken as a result of many consultation responses citing poor DAB quality, found that ninety-four per cent of DAB listeners thought DAB was at least as good as FM.

Some areas of the country are not presently covered by DAB; the BBC says that it may not be able to provide coverage to the final ten per cent of the population, and may use DRM instead. Ofcom estimate that, even after extra spectrum has been allocated to DAB, around ninety local radio stations will be unable to transmit on DAB, either because there is no space for them on a local DAB ensemble or because they cannot afford the high transmission costs of DAB that the multiplex operators are charging.

In 2005 Ofcom announced that they would be advertising for the second national digital multiplex. This enraged Gcap, the owners of the Digital One ensemble. They threatened to take the authority to court as they had been told that there wouldn't be another national multiplex. The case was dropped when Ofcom assured Gcap that stations on the proposed multiplex would not compete with stations on their ensemble.

Ofcom received two applications for the new ensemble, from National Grid Wireless and 4 Digital. The authority announced that they had awarded the licence to 4 Digital on 6 July 2007. The consortium, led by Channel 4 Radio, was a combination of existing and new commercial operators.

Under the terms of the licence being awarded, 4 Digital were required to launch services within a year. However, Channel 4 Radio withdrew from the group in October 2008. Ofcom held talks with the remaining members of the consortium but were unable to reach a satisfactory conclusion. Subsequently the licence was withdrawn.

A few years later Ofcom decided to try again. On 29 January 2015 they announced that they had received two bids. The first was from Listen2Digital, run principally by Orion Media and Babcock International Group with a few other minor members. The second bidder was Sound Digital, run by Arqiva, Bauer and UTV, amongst others. Two months later, on 27 March, Ofcom announced that Sound Digital had won the franchise. The official launch took place on 29 February 2016, with the majority of stations beginning on this date and the rest following over the next month.

The main complaint following the launch was lack of coverage over the United Kingdom. The multiplex only covered around seventy-three per cent of the population and complaints were made from Cornwall, parts of East Anglia, Scotland and Kent. In 2018 Sound Digital announced that it would add nineteen transmitters in the South West, Wales, East Anglia and the North of Scotland, increasing their coverage by nearly four million new listeners in more than 1.6 million new households.

Ofcom awarded ten trial licences in 2015 to parties in different areas who wanted to operate a small-scale DAB multiplex. The trial multiplexes covered a relatively small geographical area compared to local and national DAB multiplexes. The small-scale DAB trials kept costs low by making use of comparatively low-cost transmission equipment and freely available 'open-source' software.

Brighton was the first to launch when Ofcom engineer Rashid Mustapha installed a low-power digital radio transmitter on a Brighton rooftop and broadcast an audio track of squawking seagulls. His test successfully delivered a robust, high-quality digital radio broadcast. Since then small-scale multiplexes have been established in Norfolk, Bristol, Manchester, Portsmouth, London, Cambridge, Aldershot, Glasgow and Birmingham.

In January 2018 Ofcom launched a consultation with the aim of including community radio stations and local commercial stations on the DAB platform. They proposed that small-scale radio multiplex licences should be awarded to both commercial and not-for-profit entities, and that space would be reserved on them for community radio stations. Over 700 expressions of interest were received and the roll out of these licences is expected to start in 2020.

The UK government initially intended to migrate the vast majority of AM and FM analogue services to digital in 2015. This was subject to certain criteria being met for coverage and listening figures for DAB. There were two targets set. The first was when national DAB coverage was comparable to FM. The second requirement was that digital listening must account for fifty per cent of all radio listening. This included listening through TV and the Internet as well as DAB.

Both these objectives were reached in 2018 and the government will undertake a review of the development of digital radio. This will be to evaluate future strategies and identify potential switchover dates.

However, it's not yet been announced when this review will begin so AM and FM transmissions are likely to continue for the foreseeable future.

DAB+ was a major upgrade to DAB, which adopted a new audio format and offered supposedly stronger signals with less signal dropout. WorldDAB, the organisation responsible for defining international digital standards, introduced it in 2006. DAB+ broadcasts have launched in several countries, including Switzerland, Malta, Italy and Australia, and several other countries are also expected to launch DAB+ broadcasts over the next few years.

The UK government had previously ruled out any transition to the new format, a decision backed by the radio industry and the Department for Culture, Media and Sport. The main objection was that DAB+ was not 'backwards compatible', so older radios wouldn't be able to decode the new format.

Despite the opposition, Ofcom began testing DAB+ on the Brighton Experimental multiplex in January 2013. The BBC announced that it would undertake a trial of DAB+ in 2014. Later that year Folder Media began a four-month trial of DAB+ on the North East Wales and West Cheshire ensembles.

DAB+ was introduced officially in early 2016. Two new stations launched services on the Portsmouth trial ensemble in January of that year. A month later Sound Digital launched three full-time DAB+ services on their national multiplex. Since then a number of stations have launched on DAB+ or switched from DAB to DAB+. In March 2017 the Brighton service became the UK's first ensemble to broadcast exclusively in the DAB+ format.

The USA adopted a different digital technology to that of Europe. HD Radio, which originally stood for 'Hybrid Digital', was the method selected by the Federal Communications Commission in 2002. HD is an in-band, on-channel digital radio technology. While HD Radio does allow for an all-digital transmission mode, some AM and FM radio stations simulcast both digital and analogue audio within the same channel as well as adding new FM channels and text information.

Listeners must purchase a new receiver to receive the digital portion of the signal. In May 2018 there were more stations in the world on the air using HD Radio technology than any other digital radio technology. At that point more than 3,500 stations were broadcasting with this format.

Another way of relaying high-quality broadcasts is via satellite. Using this facility, local and national stations can reach far beyond their country's boundaries. Radio services on these platforms are generally free to air but there are some radio services that are subscription-based. These are generally digital packages of numerous channels that don't broadcast terrestrially, most notably in North America.

Some of these services, such as Music Choice or Muzak, require a fixed-location receiver and a dish antenna. Launched in 1987, Music Choice was the first digital audio broadcast service in the world. The organisation produces music-related content for sixty-five million subscribers in the United States.

Mobile satellite services, such as Sirius, XM, and Worldspace, offered listeners the chance to travel across an entire continent, listening to the same audio programming anywhere they go. Sirius was the first to begin broadcasting on 5 January 2001. Tim McGraw was the first ever artist played on satellite radio. He gave a special welcome introduction, which segued, into his song *Things Change*.

XM uses fixed-location geostationary satellites in two positions to beam their signals down to earth. Sirius uses three geosynchronous satellites in highly elliptical orbits passing over North and South America to transmit the digital streams. The Sirius signal comes from a high elevation angle in the northern part of the continent. This higher angle makes the signal less vulnerable to drop out in cities, but more likely to disappear in tunnels and other covered areas. In these cases local signal boosters are required.

Worldspace operated from Silver Spring, Maryland, with additional studios located in Washington D.C., Bangalore, Mumbai, New Delhi, and Nairobi. The company employed two satellites and broadcast sixty-two channels, thirty-eight of which were content provided by international, national and regional third parties. At its height, Worldspace had over 170,000 subscribers in Eastern and Southern Africa, the Middle East and Asia.

Subsequently, mobile satellite services have proved difficult to sustain. Worldspace filed for bankruptcy in 2008, and European operations were liquidated the following spring. Sirius and XM only survived after the companies merged in July 2008.

The smart speaker is another method of delivery for radio that has emerged recently. A smart speaker is an interactive combination of wireless speaker and voice command device with a built-in virtual assistant. These gadgets can be used to control home automation devices via your home Wi-Fi network. The owner can activate lighting, heating, home security and various electronic devices via voice activation using 'hot words'. However, these devices excel when used to relay numerous audio sources. They can stream music from services such as Spotify and Amazon Music, and play audiobooks, podcasts and radio services.

There are several different devices available but the Amazon Echo and Google Home Assistant have gained a strong foothold in this rapidly expanding market. A 2019 survey in America revealed that listening on smart speakers had doubled year-on-year, while mobile listening remained static. The survey disclosed that nearly twenty per cent of total listening hours for traditional radio streams occurred on smart speakers.

We heard previously that the early community radio experiment, which started in Britain in the early 1990s, had largely failed. With a few notable exceptions, the initial incremental stations were not able to survive because of commercial pressures. Undaunted, the community radio sector lobbied successive governments to introduce a third tier of broadcasting.

In 2002 the Radio Authority licensed fifteen 'Access Radio stations' for a trial period of one year to test the feasibility of such stations. These pilot stations targeted a wide range of minorities and groups. Angel Radio in Hampshire targeted the over-sixties, while Awaz FM was aimed at the Asian population in Glasgow. Resonance FM served the artistic community in London while Takeover Radio gave the children of Leicester their own radio station.

The Community Radio Order 2004 established the final legal framework for full-time, long-term community radio licences in the UK. Community radio services are operated on a not-for-profit basis, with community ownership and control built in to their structures. To secure a community radio licence, the applicants had to demonstrate that the proposed station would meet the demands of a target community, together with obligatory 'social gain' objectives. Social gain may take the form of a pledge to train local people in broadcasting skills, or produce programming aimed at a minority section of the local populace.

A target community could be described as any defined local area, particularly those that cannot sustain a fully commercial broadcaster.

A target community could also refer to a particular sub-community in an area, otherwise known as a 'community of interest'.

It was this section that proved useful to Radio Caroline, the old pirate broadcaster which had been broadcasting using restricted service licences since coming onshore in 1991. Caroline maintained that its listeners could be described as a 'community of interest' and began a lengthy application process for one of these licences.

Ofcom eventually agreed and awarded the station an AM community licence to broadcast to Suffolk and north Essex via a previously redundant BBC World Service transmitter mast at Orford Ness. Radio Caroline commenced broadcasting on 22 December 2017. The old pirate finally had a full-time legal outlet forty-three years after its debut broadcast.

Not everyone was happy about the proliferation of these new stations. Following pressure from the UK's Commercial Radio Companies Association, community radio stations were subject to varying funding stipulations based on a prospective station's proximity to a commercial radio broadcaster. No community radio station was permitted to raise more than fifty per cent of its operating costs from a single source, including on-air sponsorship and advertising. The remainder of operating costs had to be met through other sources such as grants, donor income, National Lottery funding or charities.

Where a community radio station lay totally within the transmission area of a commercial station with a population of 150,000 or less, no sponsorship or advertising was permitted and all funding had to come from alternative sources. In a small number of areas a community radio station might not be licensed at all. This was to protect the financial interests of smaller commercial stations. These restrictions have subsequently been relaxed and a number of community stations have been granted a broadcast licence at the discretion of Ofcom.

There have been several rounds of licensing since then. To date over 275 licences have been granted. However, it has not been an unqualified success as over forty community stations have closed down due to lack of funds or resources. These included three of the original pilot stations: Forest of Dean Radio in Gloucestershire, Sound Radio in Hackney, and Northern Visions Radio in Belfast.

The broadcast regulator Ofcom would preside over radio during a turbulent period for the industry. A financial downturn and a relaxation of the programming rules meant that a number of established radio stations disappeared. Quasi-national networks replaced these stations. Familiar names such as Beacon Radio, Leicester Sound, and Mercia FM were rebranded as Heart, Smooth or Capital.

These networks are controlled, programmed and run from one location with the facility to opt out locally for news and advertisements. These homogeneous stations tend to air just a few hours of locally originated programmes per day and then switch over to the network output. At network centre, the presenters there have the facility to record different links for each location, which are then transmitted by the individual station. The average listener could be quite oblivious to the fact that their local transmissions are in fact emanating from a studio hundreds of miles away.

A significant number of listeners were outraged that their local stations were axed and replaced by the networked output. Nevertheless, for the big radio groups there are sound economic reasons for networking programmes from a central source; the most obvious one being reduced staffing and administration costs. So far the decision seems to have paid off as audience figures have remained at the levels they were before the introduction of networking. In some cases the audience has increased.

In North America the dissemination of programming from a distant source is nothing new. The syndication of programmes has been widespread for decades. Before radio networks matured in the United States, some early radio shows were reproduced on transcription disks and mailed to individual stations.

An early example of syndication using this method was RadiOzark Enterprises, Inc. based in Springfield, Missouri. The company produced a half-hour programme called 'Sermons in Song' and distributed it to 200 stations in the 1940s. RadiOzark later produced country music shows starring Tennessee Ernie Ford, Smiley Burnette, and George Morgan.

Many syndicated radio programmes were distributed by post, although the medium changed as technology developed, going from transcription disks to vinyl records, tape recordings and CDs. Since the advent of the Internet, many stations have opted to distribute programmes via CD-quality MP3s. Nowadays, most live syndicated radio shows are distributed using satellite technology.

The Telecommunications Act of 1996, which led to significant concentration of media ownership, facilitated the rapid deployment of both existing and new syndicated programs in the late 1990s, putting syndication on par with, and eventually surpassing, the network radio format.

Radio syndication generally works the same way as television syndication, except that radio stations are usually not organised into strict affiliate-only networks. Nowadays, radio networks generally are only distributors of radio shows, and individual stations can decide which shows to carry from a wide variety of networks and independent sources.

Some examples of widely syndicated music programmes include *Rick Dees' Weekly Top 40* and the nightly request programme, *Delilah*, heard on many US stations. Syndication is particularly popular in talk radio. Most talk radio stations are free to assemble their own array of hosts like Sean Hannity, Jim Bohannon, Rush Limbaugh and Don Imus.

National Public Radio, American Public Media and Public Radio International, all supply programmes to local public radio member stations. Some radio shows are also offered on a barter system, usually at no charge to the radio station. The system is used for live programming or pre-produced programmers, and includes a mixture of ad time sold by the programme producer as well as time set aside for the radio station to sell.

American radio continues to innovate. Smart media analysts continue to pay close attention to the American radio scene as it usually provides an indication of trends to come. Where America leads the rest of the world soon follows.

The amalgamation of radio groups into larger conglomerations, as previously witnessed in America and Australasia, has now reached the United Kingdom. Nowadays two large media corporations – Bauer Media Group and Global Media Entertainment – chiefly dominate the British radio scene. Between them they have purchased nearly all of the smaller radio groups and incorporated the cluster of stations into national brands.

Global own the national station Classic FM and have rolled out a network of branded stations. These include Capital, a hit music network of eleven stations aimed young people. The twenty-one Heart stations are targeting listeners, mainly women, with adult contemporary music. Smooth Radio has a soulful, easy listening format aimed at an older audience, and the Gold network broadcast mainly on the old AM transmitters of heritage stations that have now been rebranded Heart or Capital.

Radio X, formerly XFM, broadcasts alternative rock and independent music to a number of regions and is also available nationally on digital. LBC, the London based talk station, also gained national exposure on digital. Most Global stations have brand extensions on the digital network playing niche formats: feel-good music (Heart Extra), urban and dance (Capital XTRA), plus a decade specific stations (Heart 70s and 80s).

For several years Bauer Media lagged behind Global in terms of size and spread of stations. However, in 2019 they swallowed up several smaller radio groups such as UKRD, The Wireless Group and the Lincs FM Group. It's unclear at the moment how these new stations will fit within their existing portfolio as the purchases have been referred to the Competition and Markets Authority for investigation. The new stations are being run as separate entities while the investigation is taking place.

The Bauer stations can be sorted into two main groups. The 'City' network consists of services targeting a specific geographical area on FM/AM and digital platforms. The 'Hits Radio' network, as the name implies, plays the contemporary hits of the day. The twenty-one network stations have largely kept their heritage brand names, but stations such as Northsound, Hallam, Clyde, Metro and Forth now say they are 'part of the Hits Radio network'.

There is at least seven hours of local programming on weekdays and four hours at weekends, which is mainly produced from the originating station's studios. The principal exceptions are TFM, which shares all programming with Newcastle-based Metro Radio, and the Free Radio group of stations, which produce regional programming from Birmingham. In Scotland, bespoke national programming is produced and broadcast from the studios of Clyde in Glasgow with some output from Forth in Edinburgh. Hits Radio in Manchester provides the majority of network output in Northern England and the West Midlands.

Greatest Hits Radio is a classic oldies counterpart to Hits Radio and aimed principally at a more mature audience. The eighteen stations in the network are predominantly the AM sister stations of the corresponding Hits Radio Network station. They broadcast mostly on AM and DAB, in the West Midlands, northern England and Scotland.

However, Scotland and England have two separate services, largely carrying their own programming, although some off-peak output is broadcast across both networks.

There are three FM stations incorporated within the Greatest Hits service. West Sound carries a mix of local and networked programming, including local news and travel, live sports coverage and specialist shows aimed at Dumfries and Galloway. The West Midlands outlet carries a regional three-hour drive time show each weekday afternoon and provides a weekend breakfast show to stations in northern England. Greatest Hits Liverpool serves Merseyside, Cheshire and north Wales.

The second group of Bauer stations are 'quasi-national' music-genre services broadcasting mainly through digital platforms, with some services also offered through an FM or AM outlet. Bauer have two minority stations within their inventory. Jazz FM's output of jazz, blues and soul is broadcast nationally on DAB from London. Bauer acquired this station in 2018. Scala Radio, launched in March 2019, was the first national classical music service to launch on terrestrial radio in the UK since Classic FM in 1992.

Bauer have two rock stations on their roster playing two different styles of the genre. Absolute (formerly Virgin Radio) plays rock music and also has several decade-specific digital outlets. Absolute focuses mainly on guitar-based rock while its stablemate, Planet Rock, plays classic rock on the Sound Digital multiplex.

Magic is a melodic adult-contemporary music service, available on FM in London and nationally on digital. It has also acquired three sister services: Magic Chilled, Mellow Magic and Magic Soul. Kiss plays predominantly rhythmic dance and urban music and has two sister stations. Kisstory, featuring classic dance anthems, started off as a show on the main Kiss station, but was expanded into a full-time station in 2013 in response to positive feedback from listeners. Kiss Fresh plays exclusive new releases from house, hip-hop and garage acts.

Two other stations were launched as audio equivalents of magazines in Bauer's publishing arm. Kerrang! mixed modern and classic rock with speech content targeted at a young adult audience. Heat played contemporary pop mixed with entertainment news. In 2019 both stations were removed from the digital platform, but continue on Freeview and online. Now they are mainly 'jukebox' stations with no presentation staff.

There has been an increasing consolidation of Ireland's radio industry as well, with three media companies owning the majority of Irish radio stations. Apart from RTE, Communicorp and News Corporation stations, there are only a handful of independents. However, this hasn't dented the popularity of Irish radio. According to the most recent JNLR figures, more than 3.17 million listeners tune into radio every weekday, with eighty-three per cent of all adults listening to a licensed service on any given day.

And that brings us to the present day. Radio continues to thrive despite the technological, cultural and economic shifts of the twenty-first century. Television or the Internet may have become the main way the general public consume their media, but radio continues to be a valuable asset. Radio doesn't need to drastically change its core characteristics. Its simplicity and portability will ensure it remains.

Listening to radio on additional platforms such as satellite, smart speakers, mobile phones and the Internet will steadily increase. 5G wide-area cellular mobile networks are gradually becoming available around the world and it's highly possible that 5G may replace FM and Internet streaming in the future, though broadcast radio will continue to be the predominant medium through which we enjoy radio.

What radio does need to do is continue its progress into a digital world and pay close attention to the changing pattern of consumer and listener behaviour. Radio needs to seek out new markets and continue to innovate and adapt to new technology. By adopting new broadcast platforms such as satellite, smart speakers, digital or the Internet, radio faces a multi-platform future. Radio will continue to occupy an important place in the media landscape of the twenty-first century.

Bibliography

Briggs, A., *The Birth of Broadcasting: The History of Broadcasting in the United Kingdom*, (Oxford University Press, 1961).

Briggs, S., *Those Radio Times*, (Littlehampton Book Services Ltd, 1981)

Burnett, G., *Scotland on the Air,* (Moray Press, 1938).

Buxton, F. and Owen, B., *Radio's Golden Age*, (Easton Valley Press, 1966)

Campbell, R., *The Golden Years of Broadcasting*, (Routledge Books, 1976).

Carneal, G., *Conqueror of Space: The Life of Lee de Forest*, (H Liveright, New York, 1930).

Cox, J., *American Radio Networks: A History*, (McFarland & Company, 2009).

Coe, D., *Marconi: Pioneer of Radio*, (J. Messner, 1944).

Coe, L., *Wireless Radio: A Brief History*, (McFarland & Co, 2006).

Donovan, P., *The Radio Companion*, (Grafton, 1991).

Downes, P. and Harcourt, P., *Voices in the Air: Broadcasting in New Zealand*, (Cengage Learning (EMEA) Ltd, July 1977).

Dunlap, O.E., *Marconi: The Man and His Wireless*, (Macmillan Company, 1937).

Dunlap, O.E., *Radio's 100 Men of Science,* (Harper & Brothers, New York, 1944).

Fessenden, H.M., *Fessenden: Builder of Tomorrows*, (Coward McCann, New York, 1940).

Floherty, J., *On the Air: The Story of Radio*, (Doubleday, Doran & company, 1937).

Garratt, G.R.M., *The Early History of Radio: From Faraday to Marconi,* (Institution of Engineering and Technology, 1995).

Geeves, P., *Australia: The Dawn of Broadcasting*, (Federal Publishing, 1993).

Gifford, D., *The Golden Age of Radio: An Illustrated Companion*, (Batsford Ltd, 1985).

Hennessey, B., *The Emergence of Broadcasting in Britain*, (Southerleigh, 2005).
Henslow, M., *The Miracle of Radio,* (Evans Brothers, 1946).
Horten, G., *Radio Goes To War*, (University of California Press, 2003).
Levine, I.E., *Electronics Pioneer: Lee de Forest*, (Harper Collins, 1964)
Lockhart, H., *On My Wavelength*, (Impulse Books, Aberdeen, 1973).
McDowell, W.H., *The History of BBC Broadcasting in Scotland 1923-1983*, (Edinburgh University Press, 1993).
Mulryan, P., *Radio Radio: Radio in Ireland*, (Borderline, Dublin, 1988).
Nichols, R., *Radio Luxembourg the Station of the Stars,* (Comet, London, 1983).
Street, S., *A Concise History of British Radio*, (Kelly Publications, 2002).
Street, S., *Crossing the Ether*, (John Libbey, 2006).
Skues, K., *Pop Went the Pirates: History of Offshore Radio Stations*, (Lambs' Meadow Publications, 1994).
Tomalin, N., *Daventry Calling the World*, (Caedmon, Whitby, 1998).
Took, B., *Laughter in the Air: An Informal History of British Radio Comedy*, (Robson Books, 1998).
Walker, R.R., *The Magic Spark: 50Years of Radio in Australia*, (Hawthorn Press, Melbourne, 1973).
Wallis, K., *And The World Listened: The Story of Captain Leonard Frank Plugge*, (Kelly Publications, 2008).
Wander, T., *2MT Writtle: The Birth of British Broadcasting*, (New Generation Publishing, 2010).